I Can Do Math Practice Problems GRADE 1

Volume 2: Units 5 - 8

This book belongs to:

_____

I Can Do Math Practice Problems GRADE 1

Volume 2: Units 5 - 8

Dr. Christine Scafidi

Copyright © 2025 by CKingEducation, Inc.

All rights reserved.

This publication is protected under United States and international copyright law. No part of this book may be reproduced, stored in a retrieval system, shared digitally, distributed, or transmitted in any form or by any means, electronic, mechanical, photocopying, recording, scanning, or otherwise, without the express prior written permission of the publisher. This includes, but is not limited to, uploading to shared drives, copying for multiple students, or use in multiple classrooms.

This book is licensed for use by a single teacher in a single classroom only. Reproduction or redistribution beyond the original purchase violates copyright law and publisher licensing terms.

Brief excerpts may be quoted in reviews or scholarly works in accordance with Section 107 of the U.S. Copyright Act (fair use), provided proper attribution is given.

While every effort has been made to ensure the accuracy of the content at the time of publication, the authors and publisher assume no responsibility for errors, omissions, or for any outcomes resulting from the use of this material.

Published by CKingEducation

CKingEducation

To contact CKingEducation or the authors about speaking, workshops, or ordering books in bulk, visit www.ckingeducation.com.

ISBN: 978-1-968264-03-1

Lead writer: Dr. Christine Scafidi
Editor: John Sasko
Series editor and book design: Christine King

Printed in the United States of America

Table of Contents

Track what you have done by checking off items.

I did it!	Unit/Practice/I Can Statement	Page
	Unit 5 Practice 1: I can add two-digit numbers using tens and ones.	16
	Unit 5 Practice 2: I can add two-digit numbers without making a ten.	18
	Unit 5 Practice 3: I can write an equation when adding two-digit numbers.	20
	Unit 5 Practice 4: I can add two-digit numbers to solve word problems.	22
	Unit 5 Practice 5: I can add a one-digit number to a two-digit number.	24
	Unit 5 Practice 6: I can make a ten to add one-digit and two-digit numbers.	26
	Unit 5 Practice 7: I can find the sum and write an equation.	28
	Unit 5 Practice 8: I can use base-ten drawings to show make a ten.	30
	Unit 5 Practice 9: I can add tens and ones in two-digit numbers.	32
	Unit 5 Practice 10: I can add two-digit numbers using tens and ones.	34
	Unit 5 Practice 11: I can add two-digit numbers with making a ten.	36
	Unit 5 Practice 12: I can write an equation when adding two-digit numbers.	38
	Unit 5 Practice 13: I can use models to solve addition with two-digit numbers.	40
	Unit 5 Practice 14: I can practice two-digit addition problems.	42
	Unit 5 Fluency Practice	44

Table of Contents

Track what you have done by checking off items.

I did it!	Unit/Practice/I Can Statement	Page
	Unit 6 Practice 1: I can compare lengths of objects.	50
	Unit 6 Practice 2: I compare the length of two objects to a third object.	52
	Unit 6 Practice 3: I can compare and order lengths of 3 objects.	54
	Unit 6 Practice 4: I can practice addition and subtraction using cubes.	56
	Unit 6 Practice 5: I can measure length using cubes as units.	58
	Unit 6 Practice 6: I can measure length with like units.	60
	Unit 6 Practice 7: I can measure with small and large units.	62
	Unit 6 Practice 8: I can use base-ten blocks as units to measure length.	64
	Unit 6 Practice 9: I can work with numbers to 120.	66
	Unit 6 Practice 10: I can practice addition.	68
	Unit 6 Practice 11: I can measure objects and compare lengths.	70
	Unit 6 Practice 12: I can solve story problems using length.	72
	Unit 6 Practice 13: I can solve story problems with the unknown in all positions.	74
	Unit 6 Practice 14: I can use addition and subtraction equations.	76
	Unit 6 Practice 15: I can write equations to solve story problems.	78

Table of Contents

Track what you have done by checking off items.

I did it!	Unit/Practice/I Can Statement	Page
	Unit 6 Practice 16: I can count and write numbers to 120.	80
	Unit 6 Practice 17: I can solve story problems using more or fewer.	82
	Unit 6 Fluency Practice	84
	Unit 7 Practice 1: I can name and sort solid shapes.	90
	Unit 7 Practice 2: I can put solid shapes together to make a new shape.	92
	Unit 7 Practice 3: I can name and sort flat shapes.	94
	Unit 7 Practice 4: I can draw flat shapes.	96
	Unit 7 Practice 5: I can show different triangles.	98
	Unit 7 Practice 6: I can show rectangles and squares.	100
	Unit 7 Practice 7: I can build with flat shapes.	102
	Unit 7 Practice 8: I can practice adding and subtracting using shapes.	104
	Unit 7 Practice 9: I can show equal-size pieces using shapes.	106
	Unit 7 Practice 10: I can name equal-size pieces using shapes.	108
	Unit 7 Practice 11: I can find the bigger piece or an equal-size piece.	110
	Unit 7 Practice 12: I can practice working with numbers and shapes.	112

Table of Contents

Track what you have done by checking off items.

I did it!	Unit/Practice/I Can Statement	Page
	Unit 7 Practice 13: I can tell time to the hour.	114
	Unit 7 Practice 14: I can use the hands of a clock to tell time.	116
	Unit 7 Practice 15: I can read and write time to the hour and half hour.	118
	Unit 7 Practice 16: I can draw hands on a clock face.	120
	Unit 7 Practice 17: I can practice addition and subtraction. I can draw shapes.	122
	Unit 7 Fluency Practice	124
	Unit 8 Practice 1: I can find sums to ten.	130
	Unit 8 Practice 2: I can relate addition and subtraction problems.	132
	Unit 8 Practice 3: I can add and subtract to 20.	134
	Unit 8 Practice 4: I can solve addition and subtraction story problems.	136
	Unit 8 Practice 5: I can solve change unknown story problems.	138
	Unit 8 Practice 6: I can use two equations to solve compare story problems.	140
	Unit 8 Practice 7: I can model and count collections to 120.	142
	Unit 8 Practice 8: I can count and draw models for two-digit numbers.	144
	Unit 8 Practice 9: I can solve number riddles.	146

Table of Contents

Track what you have done by checking off items.

I did it!	Unit/Practice/I Can Statement	Page
	Unit 8 Practice 10: I can use models to write number riddles.	148
	Unit 8 Fluency Practice	150

I Can Do Math Practice Problems, Grade 1 © 2025 www.ckingeducation.com

Welcome to 1st Grade!

Doing these practice problems will support you to better understand the skills and concepts you are learning.

Unit 5
Adding Within 100

In this unit we will use place value and what we know about numbers to add two-digit numbers up to 100. We will also show our thinking with equations and use different ways to solve problems.

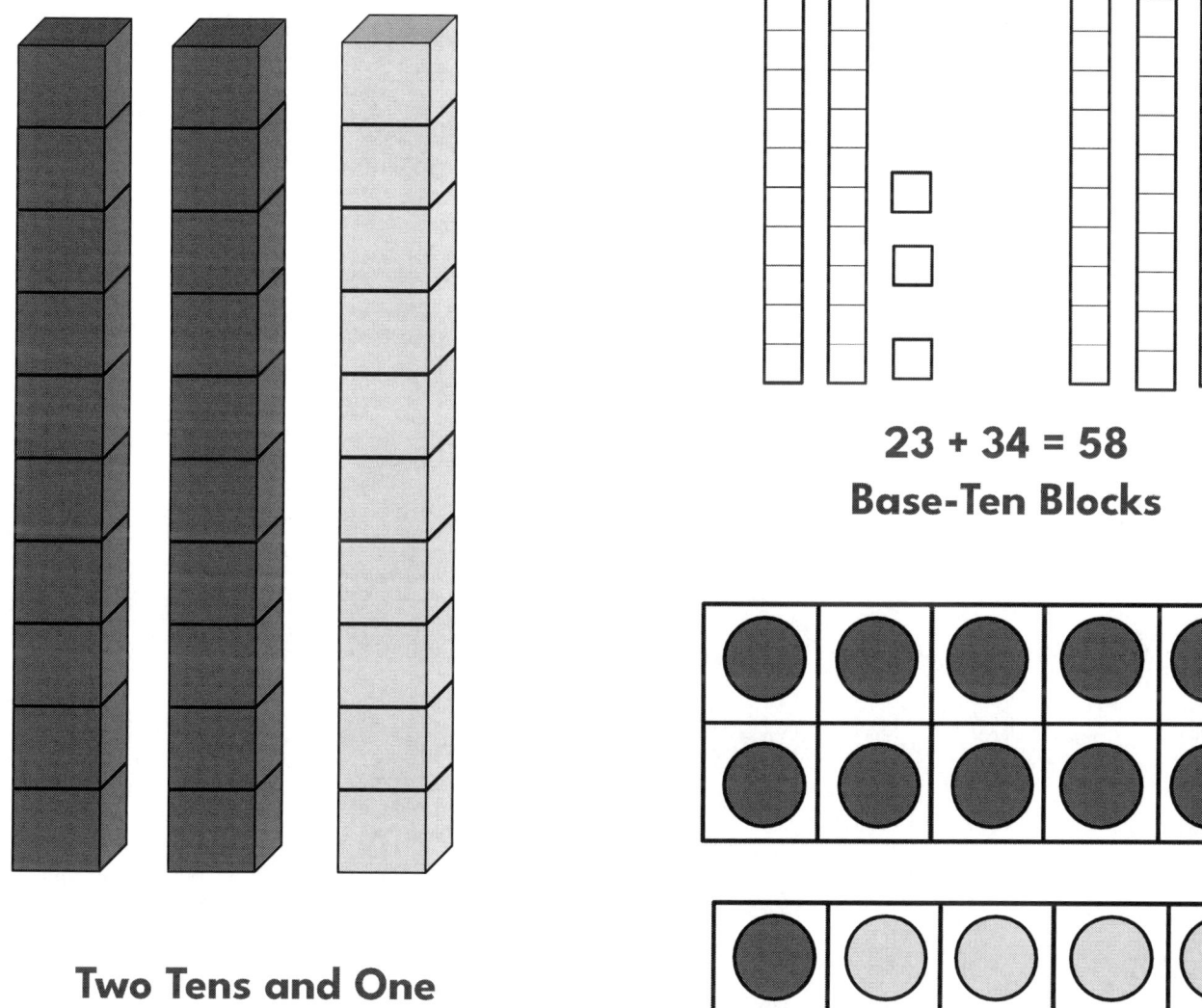

Two Tens and One More Ten is Three Tens Cubes

23 + 34 = 58
Base-Ten Blocks

11 + 9 = 20
Ten-Frames and Two Color Counters

Unit Vocabulary

Adding Within 100

Use this space to visualize the math vocabulary for this unit.

Word or Phrase	Example or Attributes	Visual Reminder

Unit Models & Strategies

Adding Within 100

Use this space to visualize the math models and strategies for this unit.

Model or Strategy	This is a...	It is used to...

Affirmation: I am math. Everything in my body in mathematical.

Name: _____ Date: _____

Unit 5 Practice 1: I can add two-digit numbers using tens and ones.

1) Add one-digit numbers.

4 + 5 = ☐   4 + 4 = ☐

8 + 1 = ☐   1 + 8 = ☐

2 + 7 = ☐   6 + 2 = ☐

3 + 4 = ☐   2 + 3 = ☐

1.OA.6

2) Add tens.

40 + 50 = ☐   40 + 40 = ☐

80 + 10 = ☐   10 + 80 = ☐

20 + 70 = ☐   60 + 20 = ☐

30 + 40 = ☐   20 + 30 = ☐

1.NBT.4

Do you understand?   ?   ✓

Affirmation: I am math. Everything in my body in mathematical.

Name: _____ Date: _____

Unit 5 Practice 1: I can add two-digit numbers using tens and ones.

3) Find the sum.

$$63 + 2$$

Equation: _____

1.NBT.4

4) Find the sum.

$$40 + 31$$

Equation: _____

1.NBT.4

Do you understand?   ?   ✓

Affirmation: I am math. Everything in my body in mathematical.

Name: _____ Date: _____

Unit 5 Practice 2: I can add two-digit numbers without making a ten.

1) Sally and Robert are using base-ten representations to add two-digit numbers.

Sally 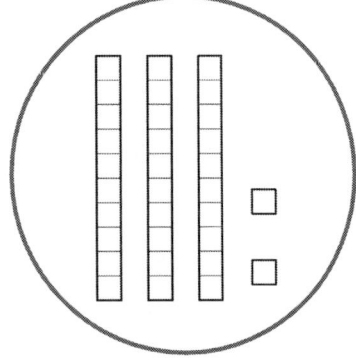   Robert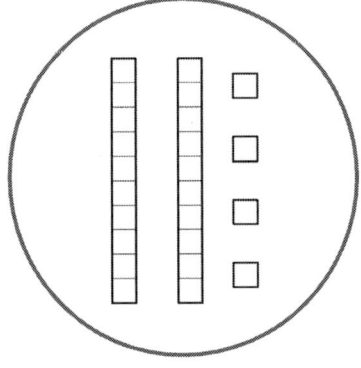

Use drawings, numbers or words to help Sally and Robert find the sum.

Equation: _____

1.NBT.4

2) Solve using base-ten representations. Use drawings, numbers or words to find the sum.

$$54 + 3$$

Equation: _____

1.NBT.4

Do you understand?   ?   ✓

Affirmation: I am math. Everything in my body in mathematical.

Name: _____   Date: _____

Unit 5 Practice 2: I can add two-digit numbers without making a ten.

3) Solve using base-ten representations. Use drawings, numbers or words to find the sum.

$$47 + 40$$

Equation: _____

1.NBT.4

4) Solve using base-ten representations. Use drawings, numbers or words to find the sum.

$$25 + 34$$

Equation: _____

1.NBT.4

Do you understand?   ?   ✓

Affirmation: I am a mathematical thinker.

Name: _____   Date: _____

Unit 5 Practice 3: I can write an equation when adding two-digit numbers.

1) **Sample:** Add tens, then add ones to solve two-digit addition.

$$34 + 12$$

Add the tens: $30 + 10 = 40$

Add the ones: $4 + 2 = 6$

Find the sum: $40 + 6 = 46$

Equation: $34 + 12 = 46$

1.NBT.4

2) Find the sum. Solve by adding the tens, then adding the ones.

$$26 + 31$$

Add the tens: _____

Add the ones: _____

Find the sum: _____

Equation: _____

1.NBT.4

Do you understand?   ?   ✓

Affirmation: I am a mathematical thinker.

Name: _____   Date: _____

Unit 5 Practice 3: I can write an equation when adding two-digit numbers.

3) **Sample:** Add on the tens, then add the ones to solve two-digit addition.

$$15 + 21$$

Add on tens: $15 + 20 = 35$

Add the ones: $35 + 1 = 36$

Equation: $15 + 21 = 36$

1.NBT.1

4) Find the sum. Solve by adding on the tens, then adding the ones.

$$52 + 24$$

Add on tens: _____

Add the ones: _____

Equation: _____

1.NBT.4

Do you understand?   ?   ✓

Affirmation: I am a mathematical thinker.

Name: _____ Date: _____

Unit 5 Practice 4: I can add two-digit numbers to solve word problems.

1) Ciara is working with pattern block shapes in the Math Center.
   She put 25 triangles on a tray.
   Then she counted out 31 squares and placed them on the tray.
   How many pattern block shapes is Ciara working with?

   Equation: _____

   1.OA.1

2) Theo is working with pattern block shapes in the Math Center.
   He has a bag with 29 pattern block shapes.
   Eighteen of the shapes are hexagons and the rest are triangles.
   How many triangles does Theo have?

   Equation: _____

   1.OA.1

Do you understand?   ?   ✓

Affirmation: I am a mathematical thinker.

Name: _____   Date: _____

Unit 5 Practice 4: I can add two-digit numbers to solve word problems.

3) Andrew has a collection of action figures.
   Some of his action figures are red.
   Thirteen of his action figures are blue.
   Andrew has a total of 37 red and blue action figures.
   How many of his action figures are red?

   Equation: _____

   1.OA.1

4) LeeAnn has a collection of 48 action figures.
   Some of her action figures are yellow and some of her action figures are green.
   How many of her action figures could be yellow and how many of her action figures could be green? Show more than one way.

   Equation: _____

   1.OA.1

Do you understand?   ?   ✓

Affirmation: I am a mathematical thinker.

Name: _____ Date: _____

Unit 5 Practice 5: I can add a one-digit number to a two-digit number.

1) Use the 10-frame to solve.

   7 + 6

   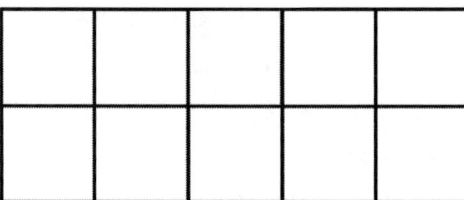

   Equation: _____

   Did you make a ten?   **YES**   **NO**

   Explain your thinking

   K.OA.4

2) Use the 10-frame to solve.

   5 + 8

   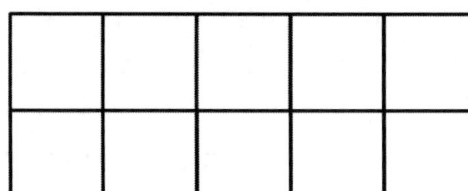

   Equation: _____

   Did you make a ten?   **YES**   **NO**

   Explain your thinking

   K.OA.4

Do you understand?   ?   ✓

Affirmation: I am a mathematical thinker.

Name: _____   Date: _____

Unit 5 Practice 5: I can add a one-digit number to a two-digit number.

3) Add 36 + 8

Equation: _____

1.NBT.4

4) Add 24 + 7

Equation: _____

1.NBT.4

Affirmation: I am a mathematical thinker.

Do you understand?   ?   ✓

Name: _____ Date: _____

Unit 5 Practice 6: I can make a ten to add one-digit and two-digit numbers.

1) Sums of ten.
   Write the number that makes each equation true.

   8 + ☐ = 10   10 = ☐ + 9

   3 + ☐ = 10   10 = ☐ + 2

   6 + ☐ = 10   10 = ☐ + 7

   5 + ☐ = 10   10 = ☐ + 1

   K.OA.4

2) Make a ten to solve.
   Use the 10-frame to model 9 + 5.

   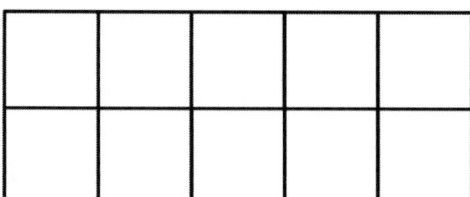

   Equation: _____

   Explain how you used make a ten to find the sum.

   K.OA.4

Do you understand?   ?   ✓

Affirmation: I am a mathematical thinker.

Name: _____ Date: _____

Unit 5 Practice 6: I can make a ten to add one-digit and two-digit numbers.

3) Find the next ten.
   How many ones are needed to make the next ten?

   25 + ☐ = 30   28 + ☐ = 30

   24 + ☐ = 30   23 + ☐ = 30

   Explain the pattern you see.

   1.NBT.4

4) Find the sum.

   57 + 8

   Equation: _____

   Did you make a ten?   **YES**   **NO**

   Explain your thinking

   1.NBT.4

Do you understand?   ?   ✓

I Can Do Math Practice Problems, Grade 1

Affirmation: I am a mathematical thinker.

Name: _____ Date: _____

Unit 5 Practice 7: I can find the sum and write an equation.

1) Circle true 👍 or false 👎.

   Use the cube towers below to solve 24 + 6.

   Add 6 ones to the cubes.

   I made a new cube tower. 👍 👎

   Equation: _____

   1.NBT.4

2) Circle true 👍 or false 👎.

   Use the cube towers below to solve 37 + 2.

   Add 2 ones to the cubes.

   I made a new cube tower. 👍 👎

   Equation: _____

   1.NBT.4

Do you understand?   ?   ✓

Affirmation: I am a mathematical thinker.

Name: _____   Date: _____

Unit 5 Practice 7: I can find the sum and write an equation.

3) Draw to solve.

$$15 + 6$$

Equation: _____

Explain how you used make a ten to solve.

1.NBT.4

4) Draw to solve.

$$5 + 38$$

Equation: _____

Explain how you used make a ten to solve.

1.NBT.4

Do you understand?   ?   ✓

Affirmation: I don't give up!

Name: _____ Date: _____

Unit 5 Practice 8: I can use base-ten drawings to show make a ten.

1) What number does this base-ten drawing show?

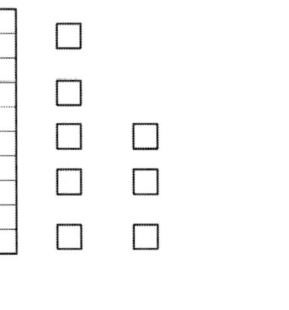

Number: _____

K.NBT.1

2) Show 47 using a base-ten drawing.

1.NBT.4

3) Find the sum. Use base-ten drawings to show your thinking.

36 + 4

Equation: _____

1.NBT.4

4) Find the sum. Use base-ten drawings to show your thinking.

36 + 8

Equation: _____

1.NBT.4

Do you understand?   ?   ✓

Affirmation: I don't give up!

Name: _____ Date: _____

Unit 5 Practice 8: I can use base-ten drawings to show make a ten.

3) The art teacher has a collection of buttons for an art project.
   There are 25 red buttons and 8 blue buttons.
   How many buttons does the art teacher have?

   Equation: _____

   1.OA.6

4) The art teacher cut out shapes for the students to use.
   He has 36 triangles and 9 large squares.
   How many shapes does the art teacher have for the students?

   Equation: _____

   1.OA.6

Do you understand?   ?   ✓

Affirmation: I don't give up!

Name: _____ Date: _____

Unit 5 Practice 9: I can add tens and ones in two-digit numbers.

1) Count on by ones.

   32, _____, _____, _____, _____, _____, _____, _____, _____

   67, _____, _____, _____, _____, _____, _____, _____, _____

   13, _____, _____, _____, _____, _____, _____, _____, _____

K.CC.1

2) Count on by tens.

   15, _____, _____, _____, _____, _____, _____, _____, _____

   34, _____, _____, _____, _____, _____, _____

   28, _____, _____, _____, _____, _____, _____, _____

1.NBT.5

Do you understand?   ?   ✓

Affirmation: I don't give up!

Name: _____ Date: _____

Unit 5 Practice 9: I can add tens and ones in two-digit numbers.

3) Find the sum.

$$37 + 14$$

Count on tens $37 + 10 = \boxed{}$

Add the ones $\boxed{} + 4 = \boxed{}$

Equation: _____

1.NBT.1

4) Find the sum.
   Finish the drawing using cube towers and cubes.

$$46 + 25$$

Equation: _____

1.NBT.4

Do you understand?   ?   ✓

Affirmation: I don't give up!

Name: _____ Date: _____

Unit 5 Practice 10: I can add two-digit numbers using tens and ones.

1) Find the sum.
   Use the ten-frames to model 17 + 15.

   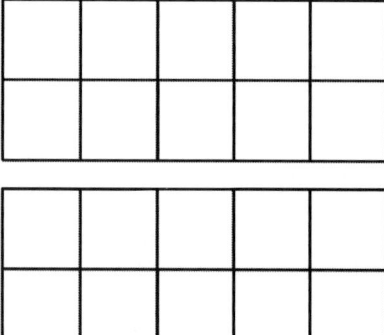 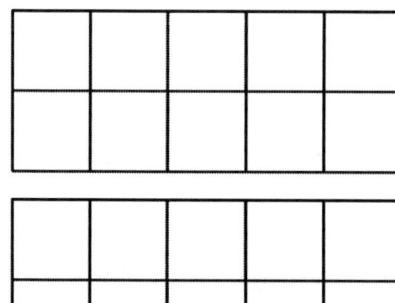

   Equation: _____

   1.OA.6

2) Find the sum.
   Finish the drawing using base-ten representations.

   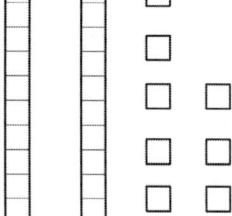

   Equation: _____

   1.NBT.4

Do you understand?   ?   ✓

I Can Do Math Practice Problems, Grade 1

Affirmation: I don't give up!

Name: _____ Date: _____

Unit 5 Practice 10: I can add two-digit numbers using tens and ones.

3) Find the sum.

   Solve by adding the tens then adding the ones.

   $$76 + 19$$

   Add the tens _____

   Add the ones _____

   Find the sum _____

   Equation: _____

   1.NBT.4

4) Find the sum.

   Solve by adding the tens then adding the ones.

   $$23 + 48$$

   Add the tens _____

   Add the ones _____

   Find the sum _____

   Equation: _____

   1.NBT.4

Do you understand?   ?   ✓

Affirmation: I don't give up!

Name: _____ Date: _____

Unit 5 Practice 11: I can add two-digit numbers with making a ten.

1) Find the sum.

$$28 + 11$$

Equation: _____

Did you make a ten?   **YES**   **NO**

Explain your thinking

1.NBT.4

2) Find the sum.

$$21 + 39$$

Equation: _____

Did you make a ten?   **YES**   **NO**

Explain your thinking

1.NBT.4

Do you understand?   **?**   **✓**

Affirmation: I don't give up!

Name: _____ Date: _____

Unit 5 Practice 11: I can add two-digit numbers with making a ten.

3) Find the sum.
   Use base-10 drawings to show your thinking.

$$46 + 18$$

Equation: _____

1.NBT.4

4) Find the sum.
   Use base-10 drawings to show your thinking.

$$12 + 39$$

Equation: _____

1.NBT.4

Do you understand?   ?   ✓

Affirmation: I don't give up!

Name: _____ Date: _____

Unit 5 Practice 12: I can write an equation when adding two-digit numbers.

1) Less than ten.
   Circle the expressions that are less than 10.

4 + 5	6 + 4	5 + 5
3 + 4	6 + 3	1 + 8
2 + 9	3 + 5	7 + 2

K.OA.4

2) Greater than or equal to ten.
   Circle the expressions that are greater than or equal to 10.

9 + 2	9 + 1	6 + 5
7 + 2	8 + 1	3 + 6
4 + 7	3 + 8	4 + 5

K.OA.4

Do you understand?   ?   ✓

I Can Do Math Practice Problems, Grade 1

Affirmation: I don't give up!

Name: _____   Date: _____

Unit 5 Practice 12: I can write an equation when adding two-digit numbers.

3) Find the next ten.
   How many ones are needed to make the next ten?

   44 + ☐ = 50   48 + ☐ = 50

   45 + ☐ = 50   41 + ☐ = 50

   47 + ☐ = 50   42 + ☐ = 50

   1.OA.8

4) Find the sum.

   19 + 76

   Equation: _____

   Did you make a ten?   **YES**   **NO**

   Explain your thinking

   1.NBT.4

Do you understand?   ?   ✓

I Can Do Math Practice Problems, Grade 1 — 39 — © 2025 www.ckingeducation.com

Affirmation: I always try, even when I am scared.

Name: _____ Date: _____

Unit 5 Practice 13: I can use models to solve addition with two-digit numbers.

1) Find the sum.
   Finish the drawing using cube towers and cubes.

   29 + 42

   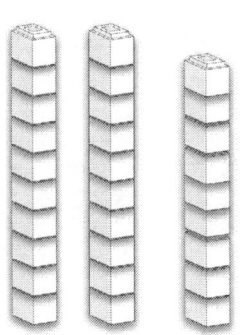

   Equation: _____

   1.OA.6

2) Find the sum.
   Finish the drawing using base-10 representations.

   44 + 38

   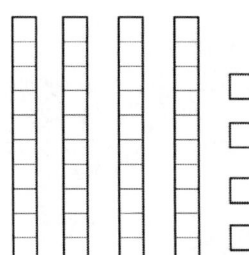

   Equation: _____

   1.OA.6

Do you understand?   ?   ✓

Affirmation: I always try, even when I am scared.

Name: _____ Date: _____

Unit 5 Practice 13: I can use models to solve addition with two-digit numbers.

3) Find the sum.

Solve by adding the tens then adding the ones.

$$17 + 64$$

Add the tens _____

Add the ones _____

Find the sum _____

Equation: _____

1.NBT.4

4) The art teacher has a star stickers for students to use in an art project. There are 36 gold stars and 44 silver stars. How many stars does the art teacher have for the students?

Equation: _____

1.OA.6

Do you understand? ? ✓

Affirmation: I always try, even when I am scared.

Name: _____   Date: _____

Unit 5 Practice 14: I can practice two-digit addition problems.

1) Paula is playing **Shake and Spill** with cubes.
She started with 52 single cubes in a can.
She spilled some cubes out of the can and made 3 cube towers and 6 single cubes.
How many single cubes are still in the can?

Equation: _____

1.OA.8

2) Larry needs 37 dog stickers for an art project.
He has 15 dog stickers.
How many more stickers does Larry need?

Equation: _____

1.OA.1

Do you understand?   ?   ✓

Affirmation: I always try, even when I am scared.

Name: _____   Date: _____

Unit 5 Practice 14: I can practice two-digit addition problems.

Questions A and B refer to the information below.

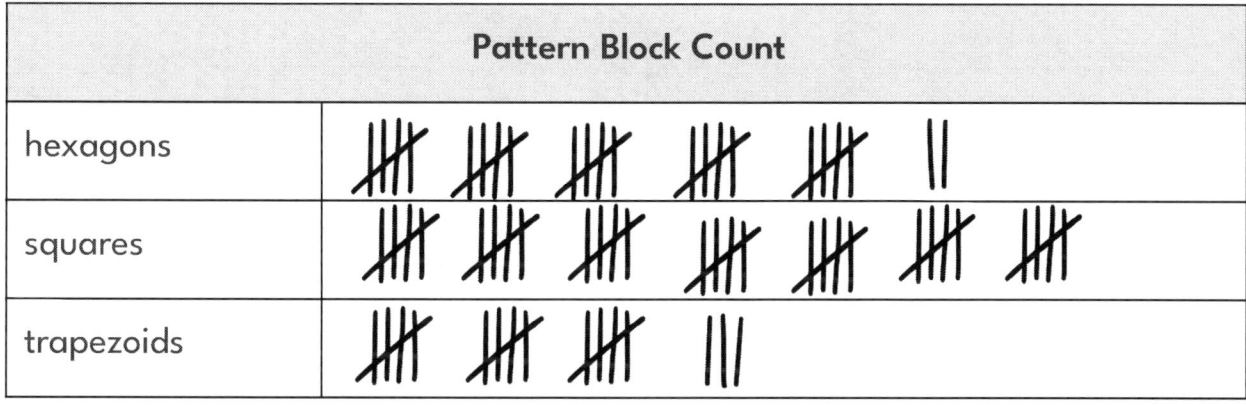

A) How many hexagons and trapezoids are in the Pattern Block Count?

Equation: _____

1.NBT.5

B) How many squares and trapezoids are in the Pattern Block Count?

Equation: _____

1.NBT.5

Do you understand?   ?   ✓

Affirmation: I get better the more I practice.

Name: _____ Date: _____

I can practice grade level fluencies.

Set 1: Add multiples of ten.

1) 24 + 10 = _____

2) 36 + 30 = _____

3) 16 + 20 = _____

4) 52 + 10 = _____

5) 52 + 20 = _____

6) 68 + 10 = _____

7) 68 + 30 = _____

1.NBT.5

Set 2: Make another ten.

1) 23 + 7 = _____

2) 46 + 4 = _____

3) 52 + 8 = _____

4) 52 + 9 = _____

5) 75 + 5 = _____

6) 65 + 5 = _____

7) 65 + 6 = _____

1.NBT.4

Do you understand?   ?   ✓

Unit Reflection

Adding Within 100

Use this space to reflect on your understanding of the unit skills and concepts.

Skill/Concept	I can...	I need to work on...

I Can Do Math Practice Problems, Grade 1

Unit 6
Length Measurements Within 120 Units

In this unit we will measure and compare the length of objects and count up to 120 using length units. We will also solve story problems by adding and subtracting, even when the missing number is in a different spot.

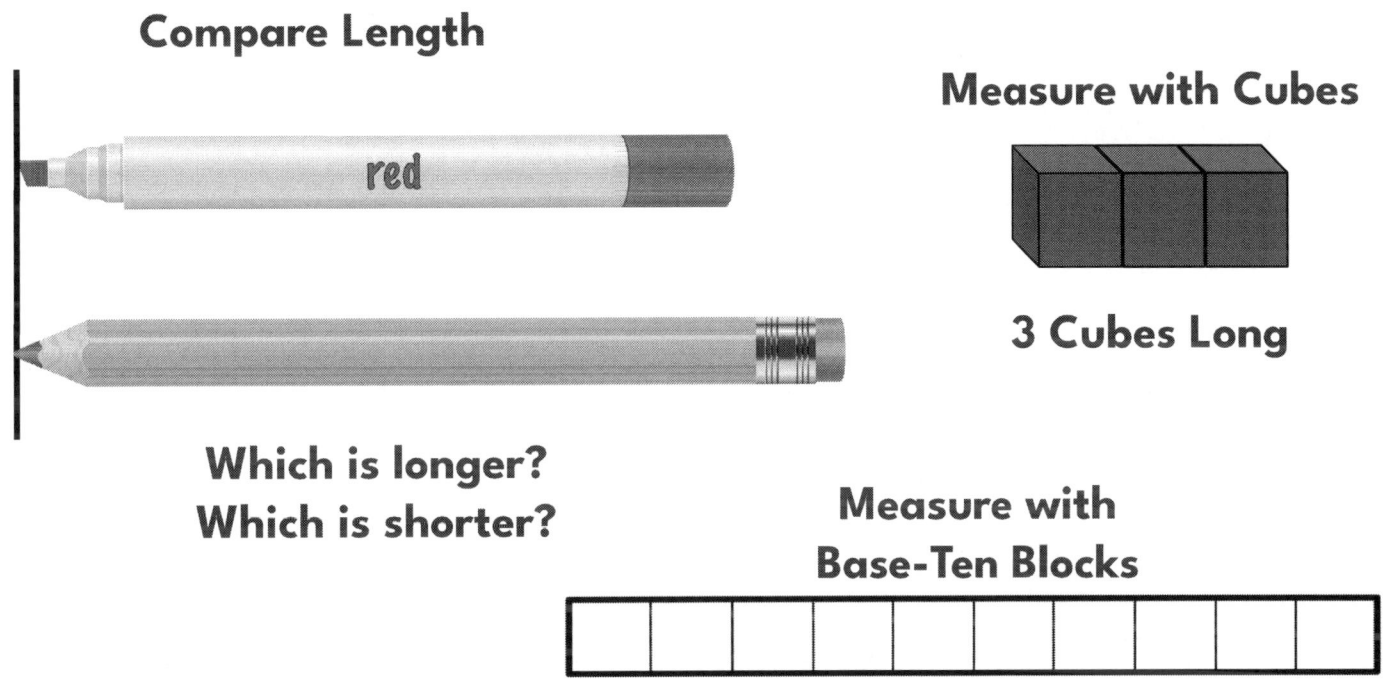

Compare Length

Which is longer?
Which is shorter?

Measure with Cubes

3 Cubes Long

Measure with Base-Ten Blocks

10 Base-Ten Blocks Long

Story Problem:
Rosie has a cube tower that is 7 units long.
She breaks off 3 units.
How long is Rosie's cube tower now?

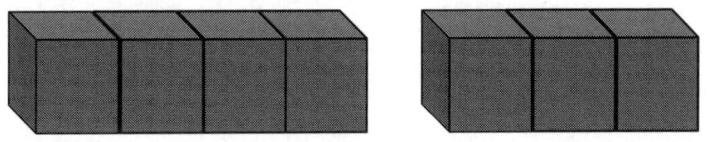

Rosie's cube tower is now 4 units long.
7 - 3 = 4

Unit Vocabulary

Length Measurements Within 120 Units

Use this space to visualize the math vocabulary for this unit.

Word or Phrase	Example or Attributes	Visual Reminder

Unit Models & Strategies

Length Measurements Within 120 Units

Use this space to visualize the math models and strategies for this unit.

Model or Strategy	This is a...	It is used to...

Affirmation: I always try, even when I am scared.

Name: _____ Date: _____

Unit 6 Practice 1: I can compare lengths of objects.

1) How many cubes in each cube tower?

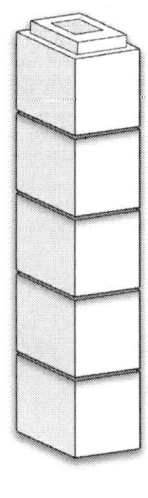

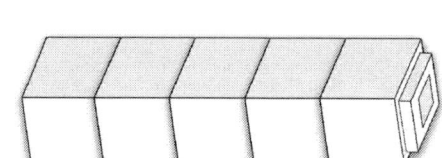

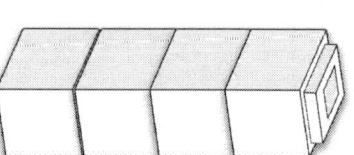

_____ cubes
Laying Down

_____ cubes
Standing Up

_____ cubes
Laying Down

K.CC.5

2) Look at the two cube towers with 3 cubes in each.

Standing Up

Laying Down

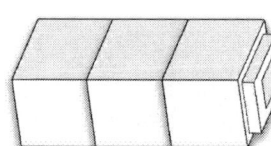

How are these cube towers the same? _____

How are these cube towers different? _____

K.CC.5

Do you understand?   ?   ✓

Affirmation: I always try, even when I am scared.

Name: _____   Date: _____

Unit 6 Practice 1: I can compare lengths of objects.

3) Look at the crayon and the marker.

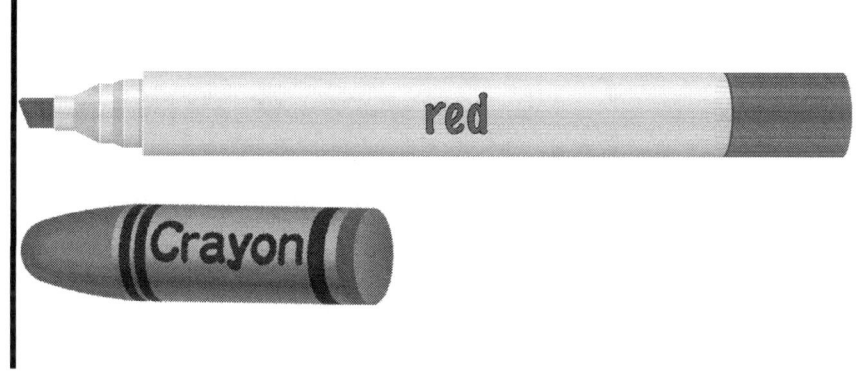

Which is longer? The _____ is longer.

Which is shorter? The _____ is shorter.

K.MD.2

4) Draw a pencil that is longer than the 4 cube tower.

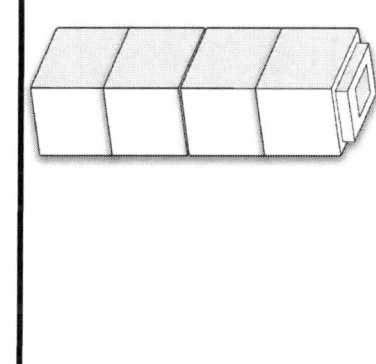

K.MD.2

Do you understand?   ?   ✓

Affirmation: I always try, even when I am scared.

Name: _____ Date: _____

Unit 6 Practice 2: I compare the length of two objects to a third object.

1) Practice addition and subtraction within 20

Find the **sum**

6 + 5 = ☐

8 + 7 = ☐

13 + 6 = ☐

9 + 8 = ☐

Find the **difference**

10 - 3 = ☐

12 - 4 = ☐

19 - 9 = ☐

15 - 5 = ☐

1.OA.6b

2) Practice finding ten more and ten less.

Find the **sum**

34 + 10 = ☐

14 + 10 = ☐

46 + 10 = ☐

55 + 10 = ☐

Find the **difference**

70 - 10 = ☐

68 - 10 = ☐

19 - 10 = ☐

25 - 10 = ☐

1.NBT.5

Do you understand?   ?   ✓

Affirmation: I always try, even when I am scared.

Name: _____ Date: _____

Unit 6 Practice 2: I compare the length of two objects to a third object.

3) Compare the length of the marker and the length of the crayon with the length of the 5 cube tower.

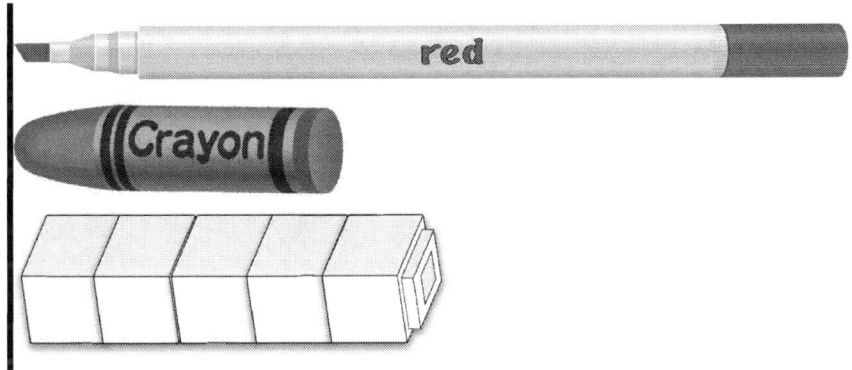

The marker is _____ than the 5 cube tower.

The crayon is _____ than the 5 cube tower.

1.MD.2

4) Noel found a toy scooter. She wants to compare the length of the toy scooter to her toy car and a paperclip.

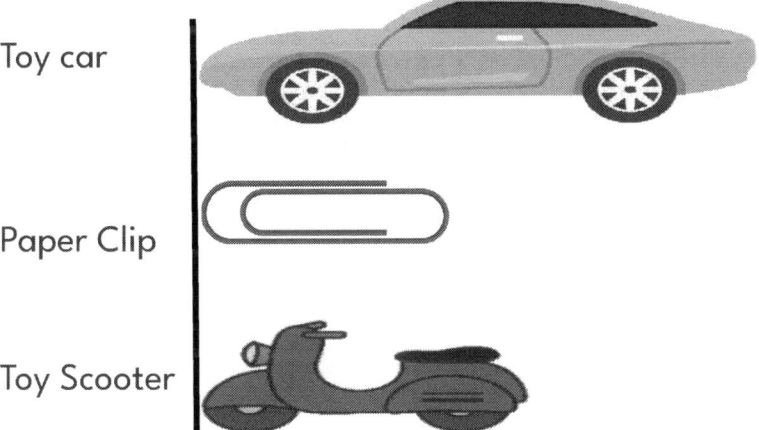

The toy scooter is _____ than the toy car and _____ than the paperclip.

1.MD.2

Do you understand?   ?   ✓

I Can Do Math Practice Problems, Grade 1

Affirmation: I always try, even when I am scared.

Name: _____ Date: _____

Unit 6 Practice 3: I can compare and order lengths of 3 objects.

1) Order the 3 objects from the shortest to the longest.

Tag

Glue Stick

Paintbrush

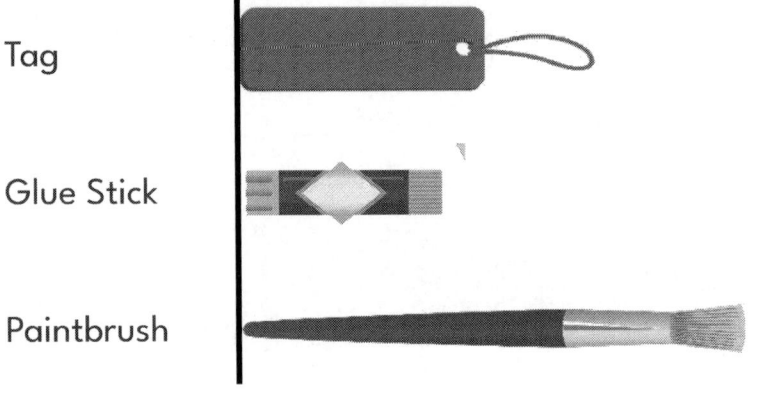

_____, _____, _____
shortest   longest

1.MD.1

2) Order the 3 objects from the longest to the shortest.

Pencil

Eraser

Marker

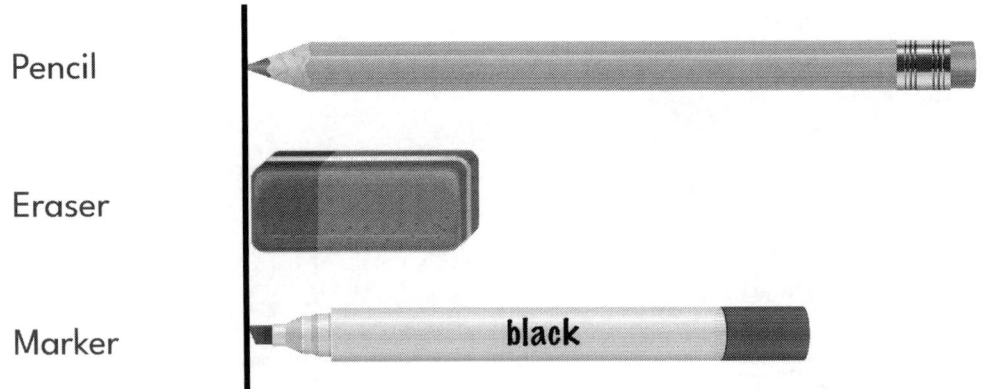

_____, _____, _____
longest   shortest

1.MD.1

Do you understand?   ?   ✓

Affirmation: I always try, even when I am scared.

Name: _____ Date: _____

Unit 6 Practice 3: I can compare and order lengths of 3 objects.

3) Which is longer?

Toy train

Toy bus

The _____ is longer than the _____.

Explain how you know: _____

_____

K.MD.2

4) Joel has a string.
   Draw an object that is shorter than the string.
   Draw an object that is longer than the string.

Joel's string

Shorter object

Longer object

1.MD.2

Do you understand?   ?   ✓

Affirmation: I always try, even when I am scared.

Name: _____ Date: _____

Unit 6 Practice 4: I can practice addition and subtraction using cubes.

1) Practice addition. Add a one-digit number to a two-digit number. Find the sum.

45 + 5 = ☐   57 + 5 = ☐

28 + 2 = ☐   62 + 9 = ☐

36 + 4 = ☐   74 + 7 = ☐

19 + 4 = ☐   85 + 9 = ☐

1.NBT.4

2) Practice finding ten more and ten less.

Find the sum.	Find the difference.
57 + 10 = ☐	80 - 10 = ☐
64 + 10 = ☐	86 - 10 = ☐
38 + 10 = ☐	37 - 10 = ☐
15 + 10 = ☐	99 - 10 = ☐

1.NBT.5

Do you understand? ? ✓

Affirmation: I always try, even when I am scared.

Name: _____ Date: _____

Unit 6 Practice 4: I can practice addition and subtraction using cubes.

3) Complete the laying down cube tower to model 7 + 5.

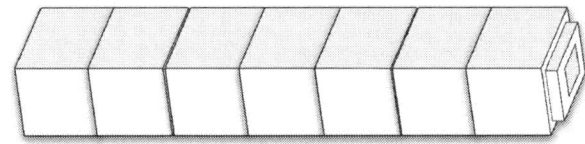

Equation: _____

1.OA.5

4) Use the laying down cube tower to model 13 - 4.

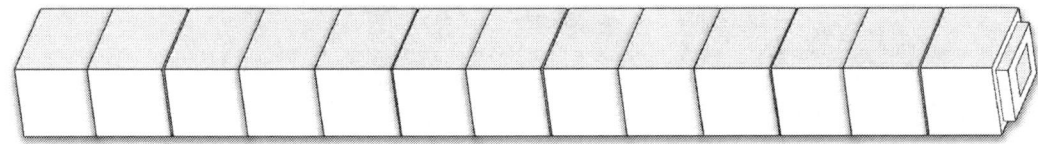

Equation: _____

1.OA.5

Do you understand?   ?   ✓

Affirmation: I always try, even when I am scared.

Name: _____   Date: _____

Unit 6 Practice 5: I can measure length using cubes as units.

1) How long is the cube tower?

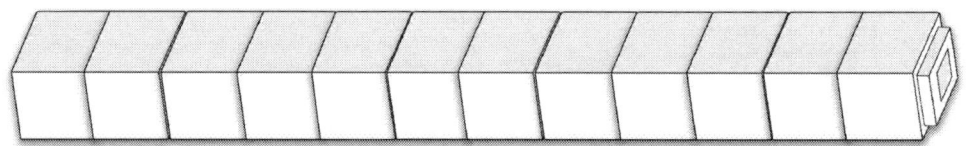

The cube tower has _____ cubes.

The cube tower is _____ cubes long.

K.CC.5

2) How long is the train?

The train is _____ cubes long.

1.MD.2

Do you understand?   ?   ✓

Affirmation: I always try, even when I am scared.

Name: _____ Date: _____

Unit 6 Practice 5: I can measure length using cubes as units.

3) Which cube tower shows the length of the marker?

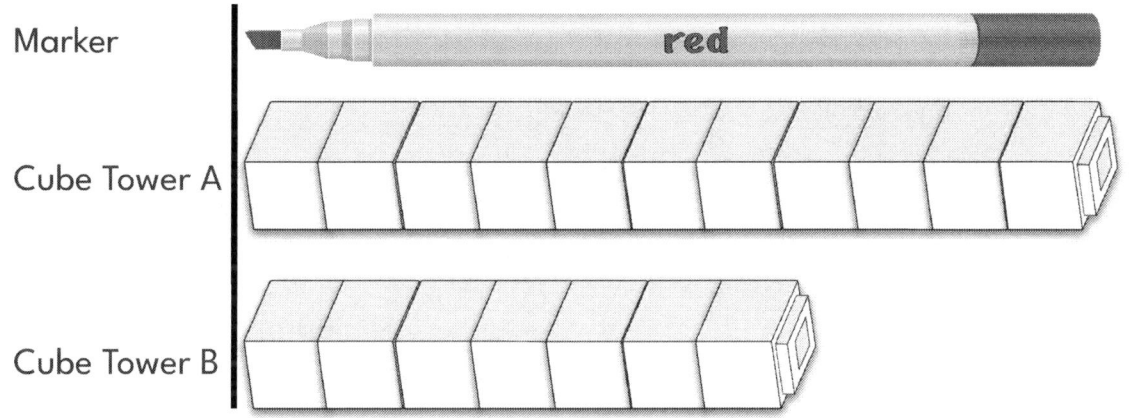

Tower _____ shows the length of the marker.

The marker is _____ cubes long.

1.MD.2

4) Same length or shorter than.

Draw a pencil that is the **same length** as the cube tower.

Draw a glue stick that is **shorter than** the cube tower.

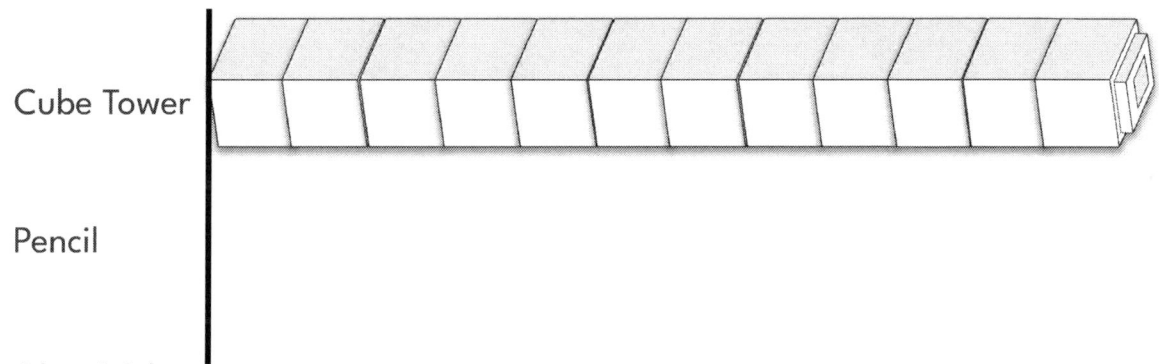

1.MD.2

Do you understand?   ?   ✓

Affirmation: I always try, even when I am scared.

Name: _____ Date: _____

Unit 6 Practice 6: I can measure length with like units.

1) Same or different.
   Which are the same? Which are different?

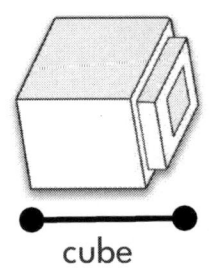
cube

quarter

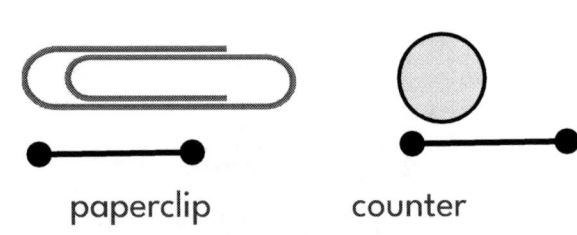
paperclip   counter

SAME	DIFFERENT

K.MD.1

2) Yes or No.
   Wilson measured a line using cubes and quarters.

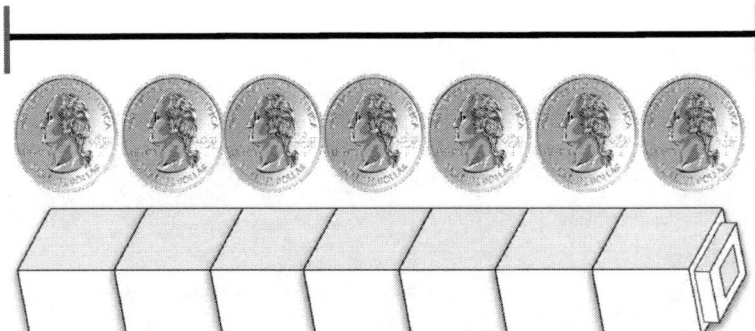

Wilson said that the cubes and quarters show the same 7 unit length.   **YES**   **NO**

Explain your thinking.

1.MD.2

Do you understand?

Affirmation: I always try, even when I am scared.

Name: _____   Date: _____

Unit 6 Practice 6: I can measure length with like units.

3) Yes or No.
   Maria measured the string using cubes as a unit.

   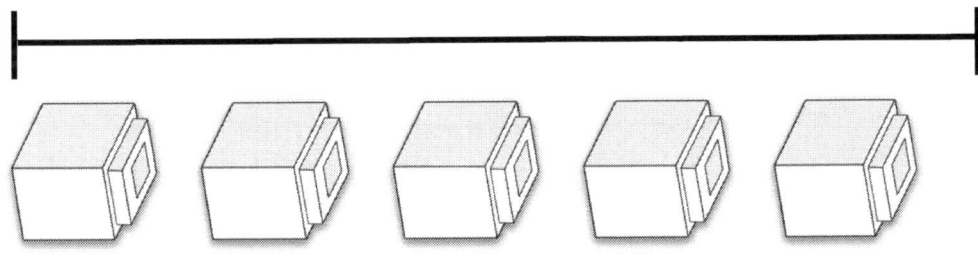

   Maria said that the string is 5 units long.   **YES**   **NO**

   Explain your thinking.

   1.MD.2

4) Yes or No.
   Amy measured the string using quarters.

   Amy said that the string is 7 units long.   **YES**   **NO**

   Explain your thinking.

   1.MD.2

Do you understand?   ?   ✓

Affirmation: I always try, even when I am scared.

Name: _____ Date: _____

Unit 6 Practice 7: I can measure with small and large units.

1) Counting base-ten blocks.

How many ones?

_____ ones

How many ones? _____ ones.

K.G.2

2) Yes or No?

Both base-ten block drawings show the count of 33. **YES**   **NO**

Explain your thinking.

K.G.2

Do you understand?   ?   ✓

I Can Do Math Practice Problems, Grade 1

Affirmation: I always try, even when I am scared.

Name: _____ Date: _____

Unit 6 Practice 7: I can measure with small and large units.

3) Measuring the same object with large units and small units.
The toy truck is measured with cubes and with base-ten blocks.

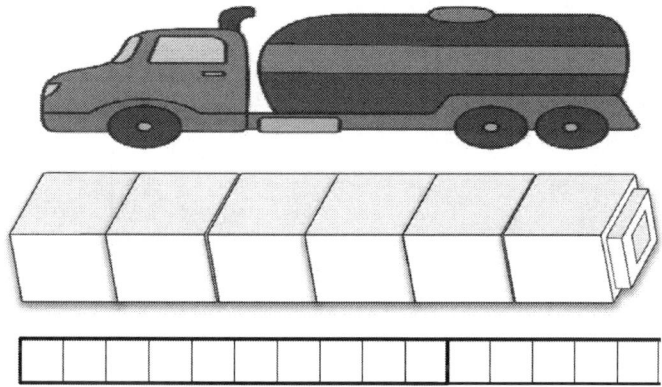

The toy truck is _____ cubes long.

The toy truck is _____ base-ten blocks long.

2.MD.2

4) **Yes or No:**
James measured the string using quarters as a unit.

James said that the string is 9 units long.   **YES**   **NO**

Explain your thinking.

2.MD.2

Do you understand?   **?**

Affirmation: I always try, even when I am scared.

Name: _____ Date: _____

Unit 6 Practice 8: I can use base-ten blocks as units to measure length.

1) Write the two-digit number for each collection of base-ten blocks.

Number: _____   Number: _____

K.CC.5

2) Count by tens.
   Use the base-ten blocks to count by ten to 100.

10, _____, _____, _____, _____, _____, _____, _____, _____, _____

K.CC.1

Do you understand?   ?   ✓

Affirmation: I always try, even when I am scared.

Name: _____ Date: _____

Unit 6 Practice 8: I can use base-ten blocks as units to measure length.

3) Michael made a design with pattern blocks.
   He used base-ten blocks as a unit to see how long his design is.

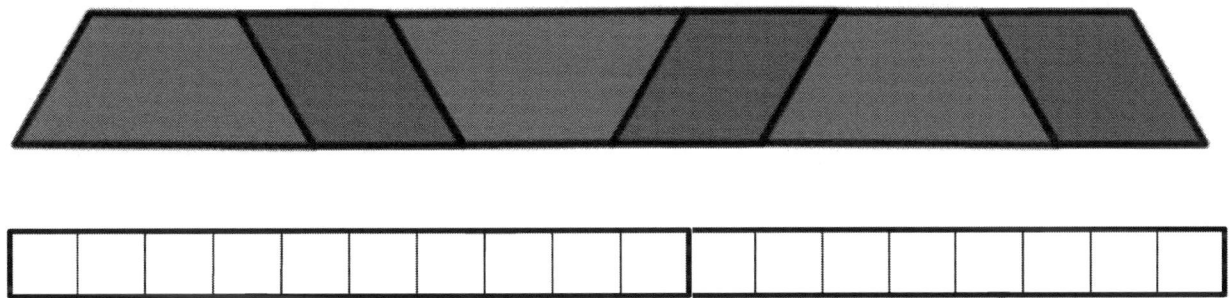

The design is _____ units long.

1.MD.2

4) Clare places 3 crayons end-to-end.
   She uses base-ten blocks as a unit to see how long the 3 crayons are.

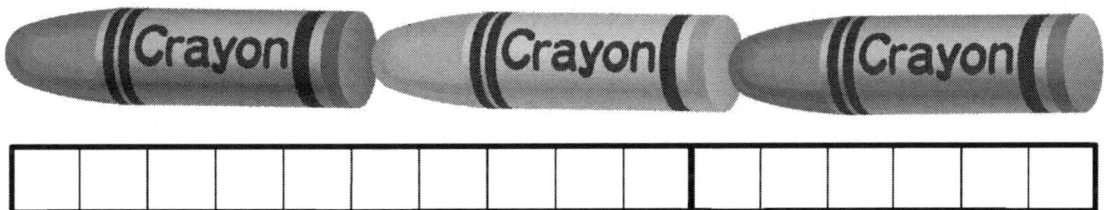

The 3 crayons placed end-to-end are _____ units long.

1.MD.2

Do you understand?   ?   ✓

Affirmation: I am brilliant, bright, and getting better every day!

Name: _____ Date: _____

Unit 6 Practice 9: I can work with numbers to 120.

1) Count on by tens to 120.

   10, _____, _____, _____, _____, _____, _____, _____, _____

   20, _____, _____, _____, _____, _____, _____, _____, _____

   40, _____, _____, _____, _____, _____, _____, _____, _____

   1.NBT.1

2) Write the number for this collection of base-ten blocks.

   Number: _____

   1.NBT.1

Do you understand?   ?   ✓

Affirmation: I am brilliant, bright, and getting better every day!

Name: _____ Date: _____

Unit 6 Practice 9: I can work with numbers to 120.

3) Use a base-ten drawing to show 98.

1.NBT.1

4) Use base-ten drawings to show 108.

1.NBT.1

Do you understand?   ?   ✓

Name: _____   Date: _____

Unit 6 Practice 10: I can practice addition.

1) Add tens to 120.

60 + 10 = ☐   60 + 40 = ☐

60 + 20 = ☐   60 + 50 = ☐

60 + 30 = ☐   60 + 60 = ☐

1.NBT.4

2) Use base-ten blocks to solve 80 + 20.

Equation: _____

1.NBT.4

Do you understand?   ?   ✓

Affirmation: I am brilliant, bright, and getting better every day!

Name: _____   Date: _____

Unit 6 Practice 10: I can practice addition.

3) Count on by ones.

93, _____, _____, _____, _____, _____, _____, _____, _____

99, _____, _____, _____, _____, _____, _____, _____, _____

105, _____, _____, _____, _____, _____, _____, _____, _____

1.NBT.1

4) Use base-ten blocks to solve 50 + 63.

Equation: _____

1.NBT.4

Do you understand?   ?   ✓

Affirmation: I am brilliant, bright, and getting better every day!

Name: _____   Date: _____

Unit 6 Practice 11: I can measure objects and compare lengths.

1) A quarter and a cube are about the same size.

   Use a quarter as a unit to measure the toy train.

   The toy train is _____ units long.

   1.MD.2

2) Use a quarter as a unit to measure your I Can Do Math Practice Problems book.

   My book is _____ units long.

   1/MD.2

Do you understand?   ?   ✓

Affirmation: I am brilliant, bright, and getting better every day!

Name: _____   Date: _____

Unit 6 Practice 11: I can measure objects and compare lengths.

3) Mike and Addie measured their toy trains with base-ten blocks as units.

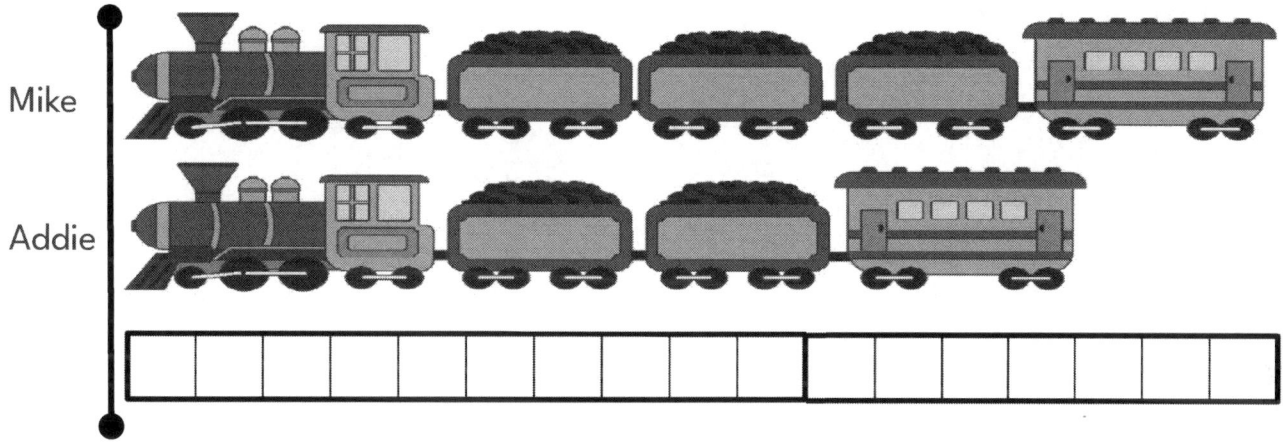

Whose toy train is longer? _____ toy train is longer.

1.MD.1

4) George measured two toy cars with base-ten blocks as units.

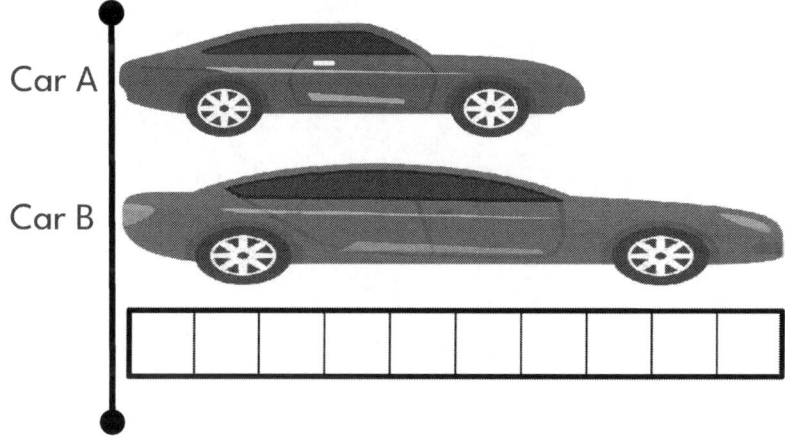

Toy Car A is _____ units long.

Toy Car B is _____ units long.

1.MD.1

Do you understand?   ?   ✓

Affirmation: I am brilliant, bright, and getting better every day!

Name: _____ Date: _____

Unit 6 Practice 12: I can solve story problems using length.

1) Bridget measured two pencils using cubes as units.
   Pencil A was 8 units long.
   Pencil B was 6 units long.
   Which pencil is longer? _____ is longer.

   How much longer?

   Equation: _____

   1.OA.1

2) Rose and Carol are measuring their pencils with cubes as units.
   Rose's pencil is 5 units long.
   Carol's pencil is 2 units longer than Rose's.
   How long is Carol's pencil?

   Equation: _____

   1.OA.1

Do you understand?   ?   ✓

Affirmation: I am brilliant, bright, and getting better every day!

Name: _____   Date: _____

Unit 6 Practice 12: I can solve story problems using length.

Questions A and B refer to the information below.

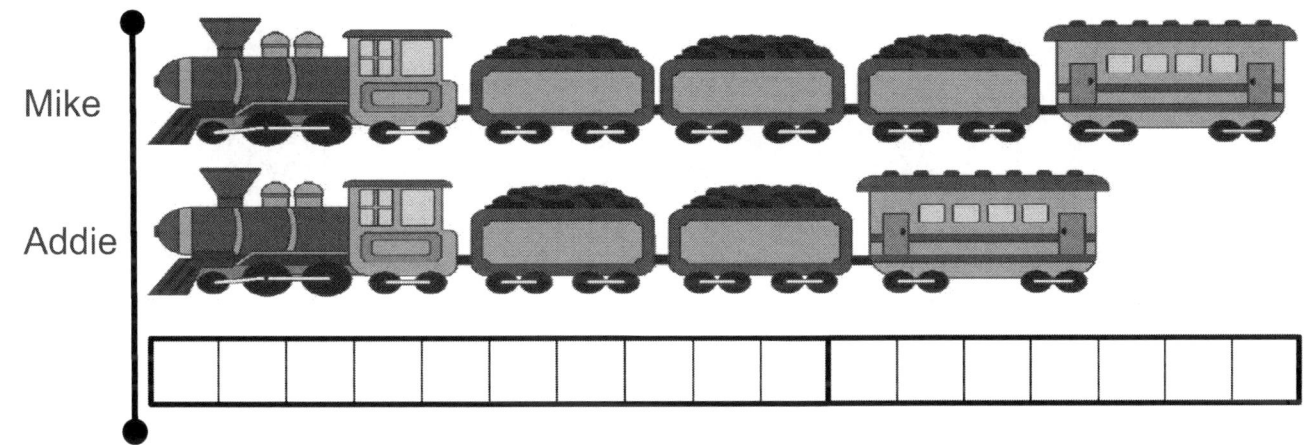

A) Mike and Addie measured their toy trains using base-ten blocks as units.

Mike's toy train is _____ units long.

Addie's toy train is _____ units long.

1.MD.2

B) How long are the toy trains together?

Equation: _____

1.OA.1

Do you understand?   ?   ✓

Affirmation: I am brilliant, bright, and getting better every day!

Name: _____ Date: _____

Unit 6 Practice 13: I can solve story problems with the unknown in all positions.

1) Match the number that makes each equation true.

16 + 14 = ☐

22 + ☐ = 42

☐ + 37 = 78

☐ + 12 = 47

41

30

35

20

1.OA.6

2) Dominick is using a piece of ribbon to wrap gifts.
   The ribbon is 16 units long.
   He cuts off a length of 6 units.
   How much ribbon is left?

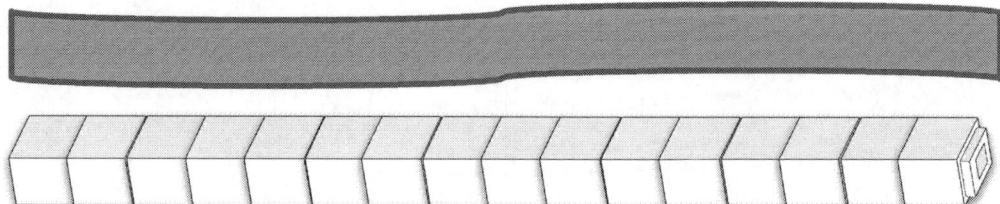

Equation: _____

1.OA.1

Do you understand?   ?   ✓

Affirmation: I am brilliant, bright, and getting better every day!

Name: _____ Date: _____

Unit 6 Practice 13: I can solve story problems with the unknown in all positions.

3) MaryAnn is using a piece of ribbon to wrap gifts.
   The ribbon is 26 units long.
   She cuts some ribbon off and now has 15 units left.
   How much ribbon did Mary cut off?

   Equation: _____

4) Rachel is using a piece of ribbon to wrap gifts.
   She cuts off 26 units.
   She now has 24 units left.
   How much ribbon did she have to start?

   Equation: _____

Do you understand?   ?   ✓

Affirmation: I love thinking! I am a thinker.

Name: _____ Date: _____

Unit 6 Practice 14: I can use addition and subtraction equations.

1) Circle true or false  .

   Circle true if both equations show related facts.

   24 - 12 = ☐   and   12 + 12 = 24   👍 👎

   16 + ☐ = 48   and   48 + 16 = 64   👍 👎

   ☐ + 15 = 36   and   36 - 15 = 11   👍 👎

   1.OA.4

2) Kim made a paper chain with 20 links.
   Sal made a paper chain with 6 fewer links than Kim's paperchain.
   How many links were in Sal's paperchain?

   Write two different equations with empty boxes to solve.

   Equation: _____

   Equation: _____

   1.OA.6

Do you understand?   ?   ✓

Affirmation: I love thinking! I am a thinker.

Name: _____ Date: _____

Unit 6 Practice 14: I can use addition and subtraction equations.

3) Nancy had a piece of rope.
   She cut off 14 units.
   The rope is now 19 units long.
   How long was the rope to start?

   Write two different equations with empty boxes to solve.

   Equation: _____

   Equation: _____

   1.OA.6

4) Harold has red string that is 17 units long.
   Paul's green string is 2 units shorter.
   How long is Paul's green string?

   Write two different equations with empty boxes to solve.

   Equation: _____

   Equation: _____

   1.OA.6

Do you understand?   ?   ✓

I Can Do Math Practice Problems, Grade 1

Affirmation: I love thinking! I am a thinker.

Name: _____   Date: _____

Unit 6 Practice 15: I can write equations to solve story problems.

1) Mateo is working with Pattern Blocks in the math center.
   He has 27 triangles and 14 squares.
   How many Pattern Block shapes is Mateo using in the math center?

   Equation: _____

   1.OA.1

2) Harper is working with Pattern Blocks in the math center.
   She is making a design with 28 shapes.
   She has 12 hexagons and the rest are squares.
   How many squares does Harper have in her design?

   Equation: _____

   1.OA.1

Do you understand?   ?   ✓

Affirmation: I love thinking! I am a thinker.

Name: _____ Date: _____

Unit 6 Practice 15: I can write equations to solve story problems.

3) Elijah has 20 more action figures than his sister.
   His sister has 13 action figures.
   How many action figures does Elijah have?

   Equation: _____

   1.OA.1

4) Sophia has a collection of action figures.
   She gives 3 action figures to her brother.
   She now has 15 action figures in her collection.
   How many action figures did Sophia have to start?

   Equation: _____

   1.OA.1

Do you understand?   ?   ✓

Name: _____ Date: _____

Unit 6 Practice 16: I can count and write numbers to 120.

1) Count on by ones.

97, _____, _____, _____, _____, _____, _____, _____, _____

101, _____, _____, _____, _____, _____, _____, _____, _____

112, _____, _____, _____, _____, _____, _____, _____, _____

1.NBT.1

2) Use a base-ten block drawing to solve.

$$87 + 32$$

Equation: _____

1.NBT.4

Do you understand?   ?   ✓

Affirmation: I love thinking! I am a thinker.

Name: _____ Date: _____

Unit 6 Practice 16: I can count and write numbers to 120.

3) Lily made a design with triangles.
   She used base-ten blocks as a unit to see how long her design was.

The design is _____ units long.

1.MD.2

4) Ava wants to count to 120 by ones.
   She has counted to 88.
   How many more numbers will she count to reach 120?

Equation: 88 + ☐ = 120

1.NBT.1

Do you understand?   ?   ✓

Affirmation: I love thinking! I am a thinker.

Name: _____   Date: _____

Unit 6 Practice 17: I can solve story problems using more or fewer.

1) Nora and Peter took a handful of base-ten blocks.

   Nora [bar with 9 squares]
   Peter [bar with 15 squares]

   How many more base-ten blocks does Peter have in his handful?

   Equation: _____

   1.OA.1

2) Mia has 4 cube towers.
   Luna has 2 cube towers and 8 cubes.
   How many fewer cubes does Luna have than Mia?

   Equation: _____

   1.OA.1

Do you understand?   ?   ✓

Affirmation: I love thinking! I am a thinker.

Name: _____ Date: _____

Unit 6 Practice 17: I can solve story problems using more or fewer.

Questions A and B refer to the information below.

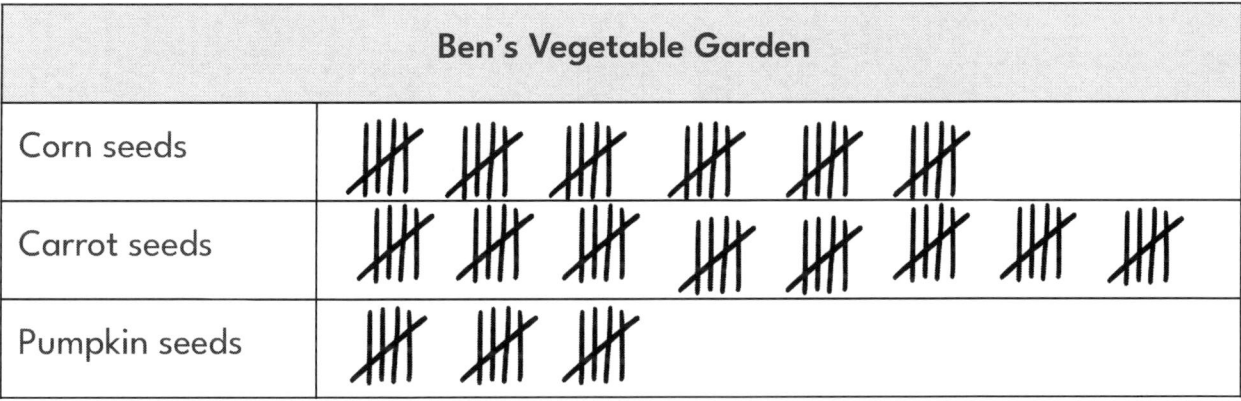

A) How many more carrot seeds than corn seeds did Ben plant?

Equation: _____

B) How many fewer pumpkin seeds than carrot seeds did Ben plant?

Equation: _____

Do you understand?   ?   ✓

Affirmation: I get better the more I practice.

Name: _____ Date: _____

I can practice grade level fluencies.

Set 1: Add within 20.

1) 9 + 8 = _____

2) 12 + 6 = _____

3) 13 + 5 = _____

4) 6 + 12 = _____

5) 2 + 17 = _____

6) 6 + 14 = _____

7) 3 + 14 = _____

1.OA.6

Set 2: Subtract within 20.

1) 20 - 10 = _____

2) 20 - 5 = _____

3) 18 - 5 = _____

4) 15 - 6 = _____

5) 20 - 7 = _____

6) 16 - 5 = _____

7) 17 - 8 = _____

1.OA.6

Do you understand?   ?   ✓

Unit Reflection

Length Measurements Within 120 Units

Use this space to reflect on your understanding of the unit skills and concepts.

Skill/Concept	I can...	I need to work on...

Unit 7
Geometry and Time

In this unit we will build and talk about shapes and their parts, and split them into equal pieces. We will also learn to tell and write time to the hour and half hour.

Solid Shapes

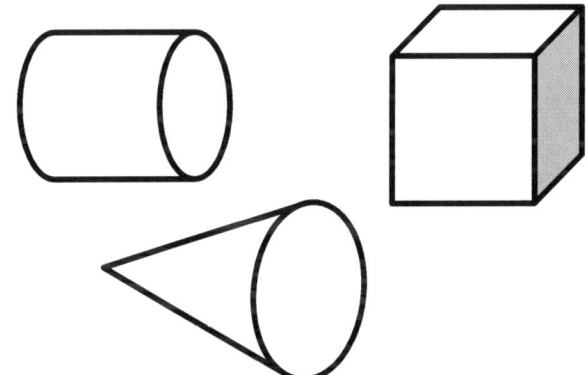

Flat Shapes

Partition the circle into 4 equal pieces

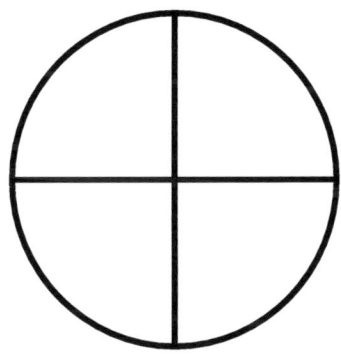

The circle is split into fourths

Analog Clock

The Clock Face Shows Eight O'Clock

Unit Vocabulary

Geometry and Time

Use this space to visualize the math vocabulary for this unit.

Word or Phrase	Example or Attributes	Visual Reminder

Unit Models & Strategies

Geometry and Time

Use this space to visualize the math models and strategies for this unit.

Model or Strategy	This is a...	It is used to...

Affirmation: I love thinking! I am a thinker.

Name: _____ Date: _____

Unit 7 Practice 1: I can name and sort solid shapes.

1) Solid or flat.
   Draw a line to match each shape to the category **Solid Shape** or Flat Shape.

 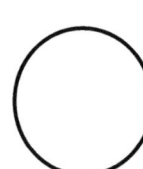    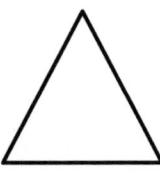

Solid Shape   Flat Shape

K.G.3

2) Match each solid shape to its name.

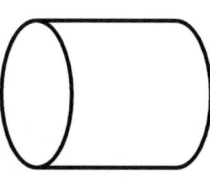  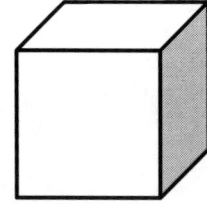 

cone   cylinder   sphere   cube

K.G.2

Do you understand?   ?   ✓

Affirmation: I love thinking! I am a thinker.

Name: _____ Date: _____

Unit 7 Practice 1: I can name and sort solid shapes.

3) Circle the solid shapes with a flat side.

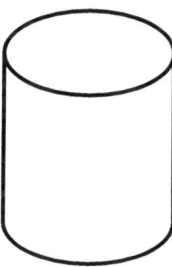  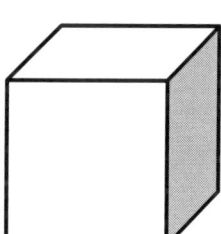 

1.G.1

4) Circle the solid shapes that roll.

 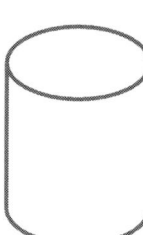 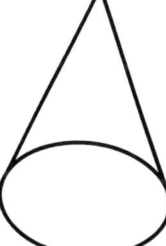 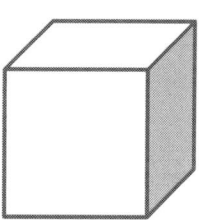

1.G.1

Do you understand?   ?   ✓

Affirmation: I am a mathematical thinker.

Name: _____ Date: _____

Unit 7 Practice 2: I can put solid shapes together to make a new shape.

1) Where am I?

Circle the position that matches where the dog is sitting.

on top of the cube   above the cube   in front of the cube

below the cube   next to the cube   behind the cube

K.G.1

2) **Agree or Disagree:**
Violet has a collection of solid shapes.
She says they are all cones.
Do you **agree** or **disagree** with Violet?

I _____ with April.

Explain your thinking.

K.G.2

Do you understand? ? ✓

Affirmation: I am a mathematical thinker.

Name: _____ Date: _____

Unit 7 Practice 2: I can put solid shapes together to make a new shape.

3) Draw a cone on **top** of the cube.
   Draw a cylinder **next to** the cube.

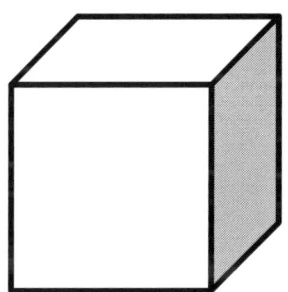

K.G.2

4) Name the solid shapes in the picture.

There are _____ and _____ in the picture.

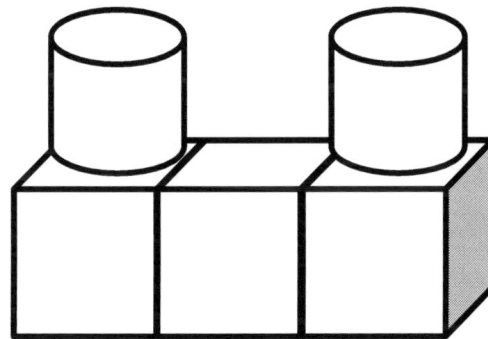

1.G.2

Do you understand?   ?   ✓

Affirmation: I am a mathematical thinker.

Name: _____ Date: _____

Unit 7 Practice 3: I can name and sort flat shapes.

1) Solid or flat.
Draw a line to match each shape to the category **Solid Shape** or **Flat Shape**.

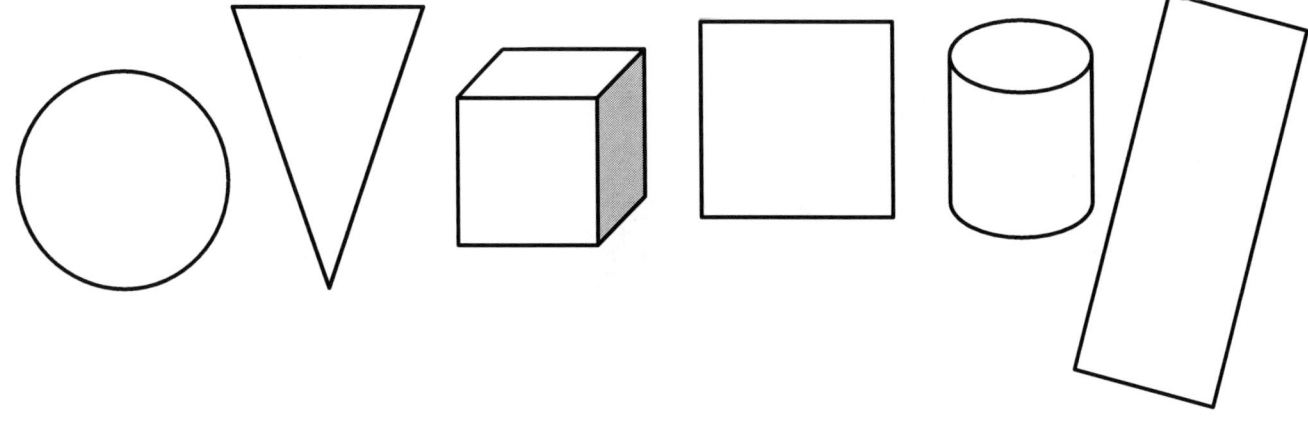

K.G.3

2) Match each flat shape to its name.

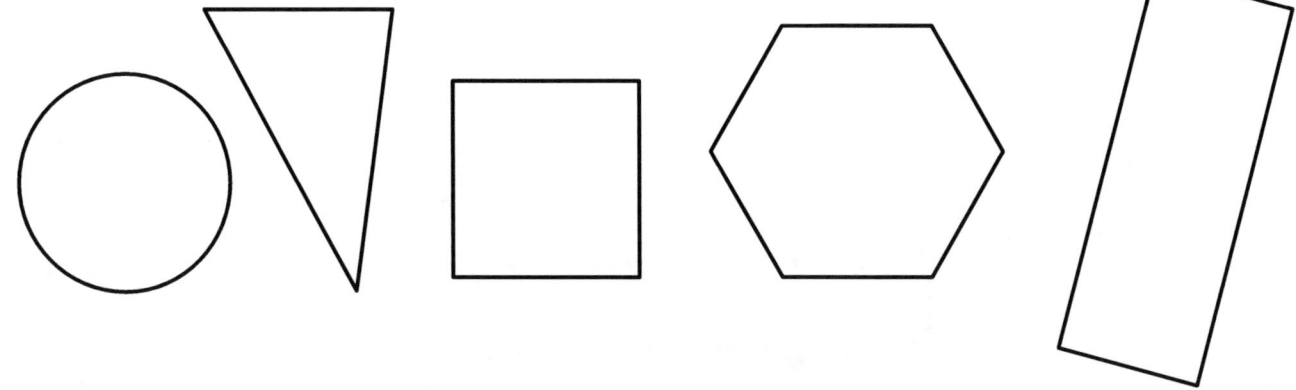

K.G.2

Do you understand?   ?   ✓

Affirmation: I am a mathematical thinker.

Name: _____ Date: _____

Unit 7 Practice 3: I can name and sort flat shapes.

3) **Same or different:**

Describe how a triangle and a square are the same.
Describe how a triangle and a square are different.

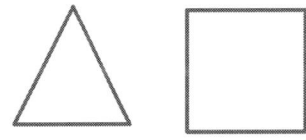

SAME	DIFFERENT
The triangle and the square are the same...	The triangle and the square are different...

1.G.1

4) **Agree or Disagree:**

Ezra has a collection of flat shapes.
He says they are not all triangles.
Do you **agree** or **disagree** with Ezra?

I _____ with Ezra.

Explain your thinking.

1.G.1

Do you understand?   ?   ✓

Affirmation: I am a mathematical thinker.

Name: _____   Date: _____

Unit 7 Practice 4: I can draw flat shapes.

1) Draw a small rectangle and a large rectangle on the dot grid.

1.G.1

2) Draw a triangle and a square on the dot grid.

1.G.1

Do you understand?   ?   ✓

Affirmation: I am a mathematical thinker.

Name: _____ Date: _____

Unit 7 Practice 4: I can draw flat shapes.

3) Madison built 3 squares using popsticks.
   How many popsticks would Madison have used?

Equation: _____

1.OA.2

4) **Large or small:**

Draw a circle that is larger.
Draw a circle that is smaller.

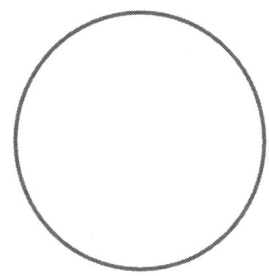

1.G.1

Do you understand?   ?   ✓

Affirmation: I am a mathematical thinker.

Name: _____ Date: _____

Unit 7 Practice 5: I can show different triangles.

1) Circle true 👍 or false 👎.

   Each shape is a triangle.

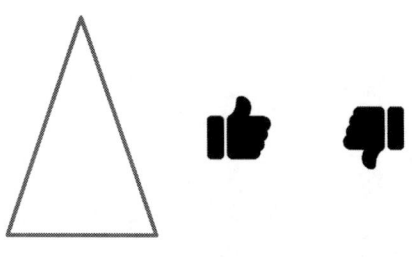

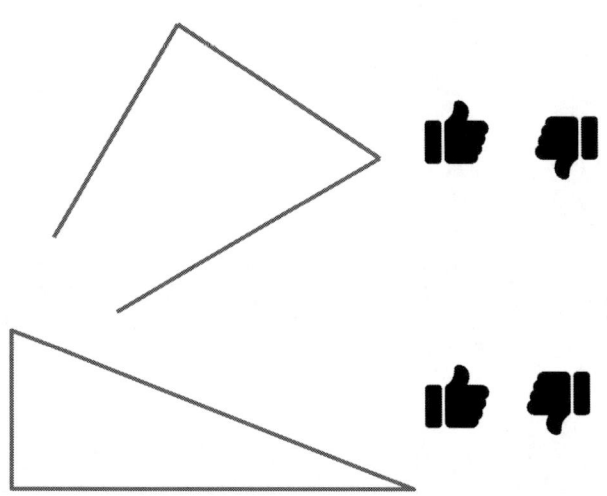

1.G.1

2) Make a chart to show triangles and not triangles.
   Draw at least 3 examples of each.

Triangles	Not Triangles

1.G.1

Do you understand?   ?   ✓

Affirmation: I am a mathematical thinker.

Name: _____   Date: _____

Unit 7 Practice 5: I can show different triangles.

3) Finish each to make a collection of triangles.

 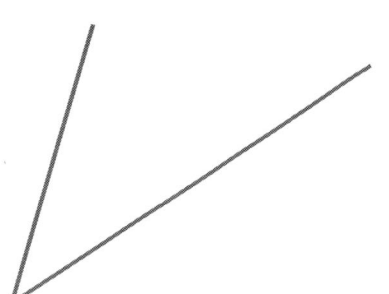

1.G.1

4) Logan said his pizza slice is a triangle.
    Ivy said the pizza slice was not a triangle.
    Do you **agree** with Logan or with Ivy?

I agree with_____.

Explain your thinking.

1.G.1

Do you understand?   ?   ✓

Affirmation: I am a mathematical thinker.

Name: _____ Date: _____

Unit 7 Practice 6: I can show rectangles and squares.

1) Find the sum.

45 + 10 = ☐   34 + 6 = ☐

45 + 20 = ☐   57 + 3 = ☐

76 + 10 = ☐   81 + 9 = ☐

76 + 20 = ☐   5 + 15 = ☐

1.NBT.4

2) Circle true 👍 or false 👎.

Each shape is a rectangle.

Do you understand?   ?   ✓

Affirmation: I am a mathematical thinker.

Name: _____   Date: _____

Unit 7 Practice 6: I can show rectangles and squares.

3) Draw 2 squares and 2 rectangles on the dot grid.

4) Which one is a rectangle?
   Dylan and Zoey were looking at a playing card and a dollar bill.
   Dylan said the playing card looks like a rectangle.
   Zoey said the dollar bill looks like a rectangle.
   Do you **agree** with Dylan or with Zoey?

I agree with_____.

Explain your thinking.

Do you understand?   ?

Affirmation: I don't give up!

Name: _____ Date: _____

Unit 7 Practice 7: I can build with flat shapes.

1) Circle the names of the flat shapes used in the pattern block design.

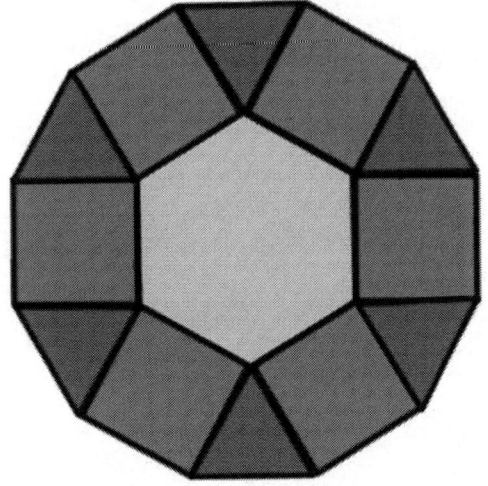

Square

Rectangle

Triangle

Circle

Hexagon

K.G.2

2) Draw these flat shapes around the square.

Draw a rectangle next to the square.
Draw a triangle above the square.
Draw a circle below the square.

K.G.1

Do you understand?   ?   ✓

I Can Do Math Practice Problems, Grade 1

Affirmation: I don't give up!

Name: _____ Date: _____

Unit 7 Practice 7: I can build with flat shapes.

3) Rachel put 2 squares together.
   What new shape did she make?

   Rachel made a _____.

   Show the new shape Rachel made.

1.G.2

4) How many of each flat shape was used in this design?

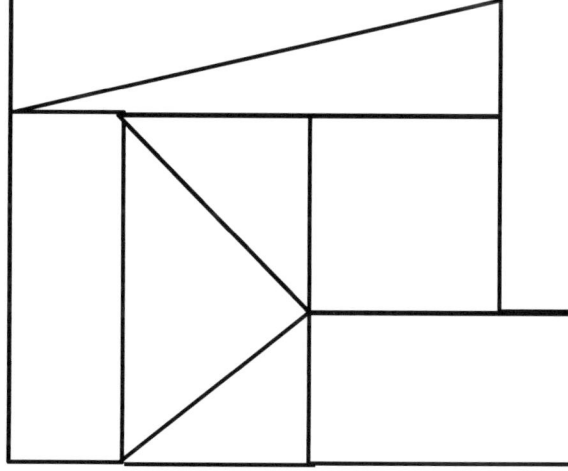

_____ rectangles

_____ triangles

_____ squares

1.G.2

Do you understand?   ?   ✓

Affirmation: I don't give up!

Name: _____ Date: _____

Unit 7 Practice 8: I can practice adding and subtracting using shapes.

1) Count the sides.
   How many sides does each shape have?

   hexagon 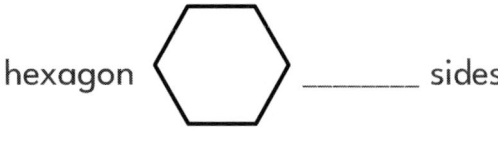 _____ sides   circle  _____ sides

   rectangle 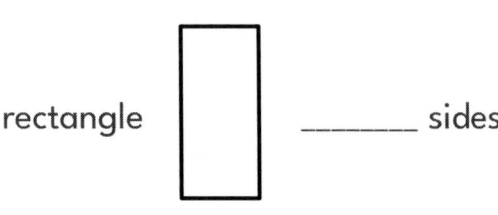 _____ sides   triangle  _____ sides

   square  _____ sides   trapezoid  _____ sides

   1.G.1

2) Add the sides.
   Andres drew 1 triangle, 1 hexagon and 1 rectangle.
   Show the shapes Andres drew.

   How many sides are in his drawing?

   Equation: _____

   1.OA.2

Do you understand?   ?   ✓

Affirmation: I don't give up!

Name: _____   Date: _____

Unit 7 Practice 8: I can practice adding and subtracting using shapes.

3) How many triangles are inside the 2 hexagons?

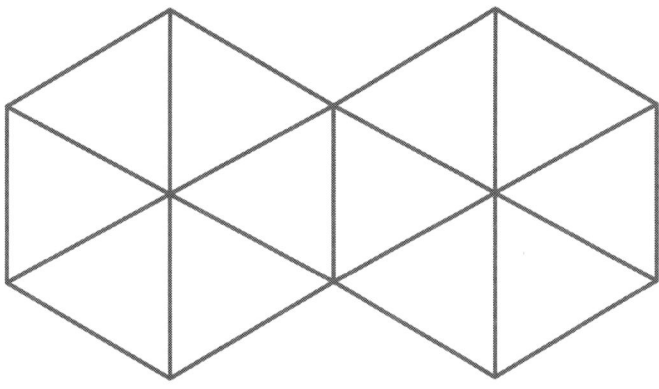

Equation: _____

1.OA.6

4) How many of each flat shape was used in this design?

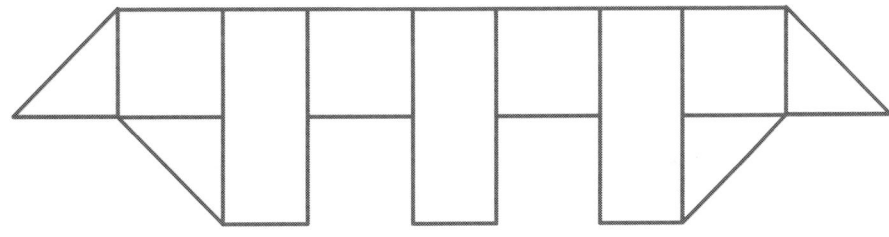

_____ rectangles   _____ triangle   _____ squares

How many total shapes?

Equation: _____

1.OA.2

Do you understand?   ?   ✓

Affirmation: I don't give up!

Name: _____ Date: _____

Unit 7 Practice 9: I can show equal-size pieces using shapes.

1) Draw a small circle and a large circle.
   Draw a small square and a large square.

1.G.1

2) Circle true 👍 or false 👎.

Each shape is split into equal-size pieces.

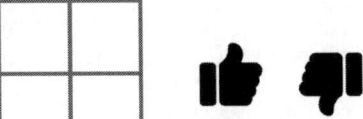

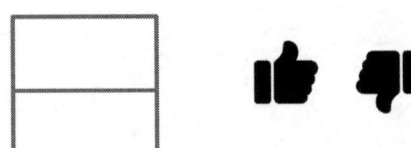

1.G.3

Do you understand?   ?   ✓

Affirmation: I don't give up!

Name: _____ Date: _____

Unit 7 Practice 9: I can show equal-size pieces using shapes.

3) Split each shape into **halves**.

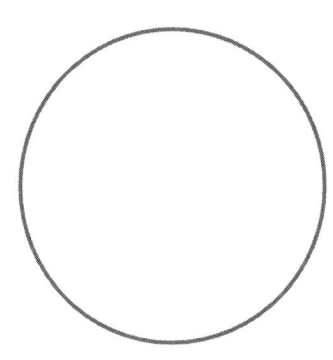  

1.G.3

4) Split each shape into **fourths**.

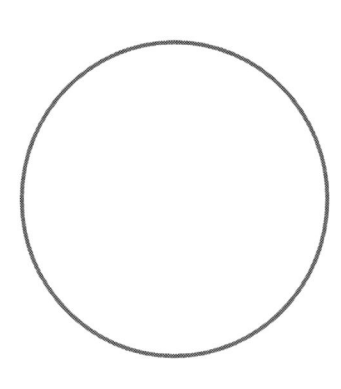  

1.G.3

Do you understand?   ?   ✓

Affirmation: I don't give up!

Name: _____ Date: _____

Unit 7 Practice 10: I can name equal-size pieces using shapes.

1) Split the shape into 2 equal pieces. Shade one of the pieces.

I shaded _____ of the circle.

1.G.3

2) Split the shape into 4 equal pieces. Shade one of the pieces.

I shaded _____ of the circle.

1.G.3

Do you understand?   ?   ✓

Affirmation: I don't give up!

Name: _____ Date: _____

Unit 7 Practice 10: I can name equal-size pieces using shapes.

3) Draw a line to match each shaded piece to the fraction.

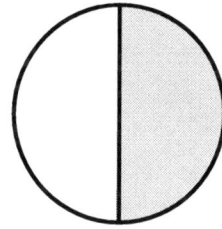   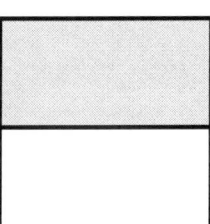   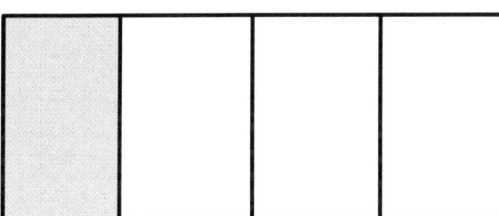

| half of | quarter of |

1.G.3

4) Hazel says both circles are split into 4 pieces.
   So, each circle shows fourths.
   Do you **agree** or **disagree** with Hazel?

   I _____ with Hazel.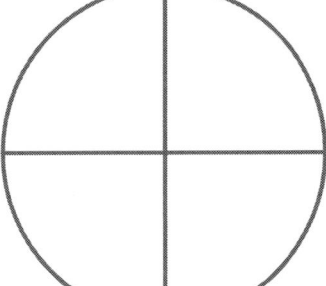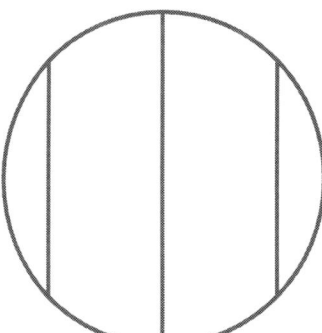

   Explain your thinking.

1.G.3

Do you understand?   ?   ✓

Affirmation: I don't give up!

Name: _____ Date: _____

Unit 7 Practice 11: I can find the bigger piece or an equal-size piece.

1) Count on by ones.

99, _____, _____, _____, _____, _____, _____, _____, _____

106, _____, _____, _____, _____, _____, _____, _____, _____

110, _____, _____, _____, _____, _____, _____, _____, _____

1.NBT.1

2) Add ten.
   Write the number that makes each equation true.

98 + 10 = ☐   10 + 100 = ☐

10 + 64 = ☐   96 + 10 = ☐

101 + 10 = ☐   10 + 110 = ☐

105 + 10 = ☐   75 + 10 = ☐

1.NBT.5

Do you understand?   ?   ✓

Affirmation: I don't give up!

Name: _____ Date: _____

Unit 7 Practice 11: I can find the bigger piece or an equal-size piece.

3) Shade one part of each rectangle.

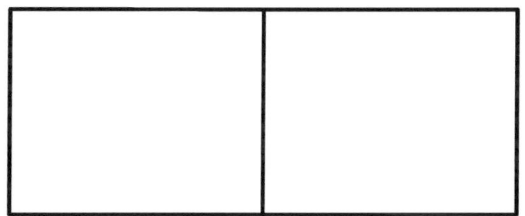

Shade one-half

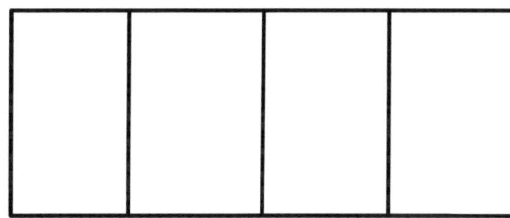

Shade one-fourth

Is one-half or one-fourth **smaller**?

_____ is smaller.

1.G.3

4) Mr. Joseph wants to share the watermelon with two lunch tables. How can he split the watermelon so that each table gets an equal piece?

Explain your thinking.

1.G.3

Do you understand?   ?   ✓

I Can Do Math Practice Problems, Grade 1

Affirmation: I always try, even when I am scared.

Name: _____   Date: _____

Unit 7 Practice 12: I can practice working with numbers and shapes.

1) Addition to 100.
   Match the number that makes each equation true.

   40 + ☐ = 100   50

   43 + ☐ = 100   57

   50 + ☐ = 100   62

   38 + ☐ = 100   60

   1.OA.8

2) Practice subtraction within 20.   Write the number that makes each equation true.

   10 - 3 = ☐   9 - ☐ = 6

   8 - 4 = ☐   7 - ☐ = 2

   12 - 10 = ☐   4 - ☐ = 3

   6 - 1 = ☐   13 - ☐ = 10

   1.OA.8

Do you understand?   ?   ✓

Affirmation: I always try, even when I am scared.

Name: _____   Date: _____

Unit 7 Practice 12: I can practice working with numbers and shapes.

3) Draw one square, one rectangle and one triangle on the dot grid below. Split each shape into 2 equal pieces.

1.G.3

4) Owen wants to put 4 squares together. What new shape could he make?

Owen could make a _____.

Show the new shape Owen could make.

1.G.2

Do you understand?   ?   ✓

Affirmation: I always try, even when I am scared.

Name: _____ Date: _____

Unit 7 Practice 13: I can tell time to the hour.

1) **Same or different:**

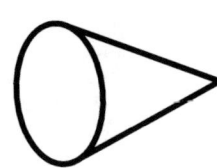  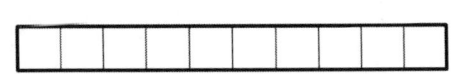 

Which objects are the same? Which object is different?

SAME	DIFFERENT

K.MD.3

2) Match the time on the analog clock to the hour it shows.

9 o'clock   7 o'clock   10 o'clock

1.MD.3

Do you understand?   ?   ✓

I Can Do Math Practice Problems, Grade 1

Affirmation: I always try, even when I am scared.

Name: _____ Date: _____

Unit 7 Practice 13: I can tell time to the hour.

3) Draw hands on each clock face to show the hour.

4 o'clock

2:00

1.MD.3

4) Maya says the hour-hand-only clock shows 8 o'clock.
   Isaac says the clock needs two hands to show the time.
   Do you **agree** with Maya or Isaac?

   I agree with _____.

   Explain your thinking.

1.MD.3

Do you understand? ? ✓

Affirmation: I always try, even when I am scared.

Name: _____ Date: _____

Unit 7 Practice 14: I can use the hands of a clock to tell time.

1) What is between?
   Circle the shape that is between.

  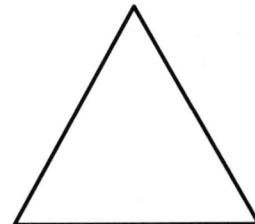

square   rectangle   triangle

The _____ is between the _____ and the _____.

K.G.1

2) Between the numbers

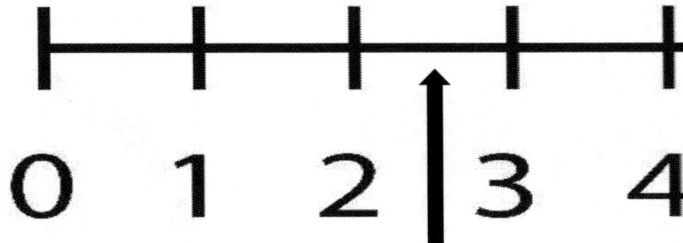

The arrow is pointing between the number _____ and the number _____.

K.G.1

Do you understand?   ?   ✓

I Can Do Math Practice Problems, Grade 1

Affirmation: I always try, even when I am scared.

Name: _____ Date: _____

Unit 7 Practice 14: I can use the hands of a clock to tell time.

3) Match the time to show if it is **half-past** or **o'clock**.

half-past   o'clock

1.MD.3

4) **Agree or Disagree:**
Zoey says the hour-hand-only clock shows half past 1.
Do you **agree** or **disagree** with Zoey?

I _____ with Zoey.

Explain your thinking.

1.MD.3

Do you understand?   ?   ✓

Affirmation: I always try, even when I am scared.

Name: _____ Date: _____

Unit 7 Practice 15: I can read and write time to the hour and half hour.

1) Add to 30.
   Match the number that makes each equation true.

   20 + ☐ = 30

   15 + ☐ = 30

   30 = ☐ + 25

   30 = 10 + ☐

   | 5 |
   | 10 |
   | 20 |
   | 15 |

   1.NBT.1

2) Half of the shape.
   Circle the pictures that show half of the shape.

   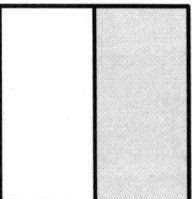 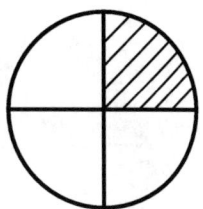 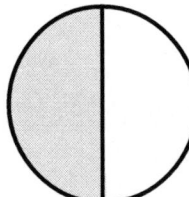 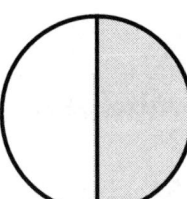 

   1.G.3

Do you understand?   ?   ✓

Affirmation: I always try, even when I am scared.

Name: _____ Date: _____

Unit 7 Practice 15: I can read and write time to the hour and half hour.

3) Write the time below each clock face.

☐ : ☐   ☐ : ☐   ☐ : ☐

1.MD.3

4) Write the time below each clock face.

☐ : ☐   ☐ : ☐   ☐ : ☐

1.MD.3

Do you understand? ? ✓

Affirmation: I always try, even when I am scared.

Name: _____ Date: _____

Unit 7 Practice 16: I can draw hands on a clock face.

1) Count on by ones.

15, _____, _____, _____, _____, _____, _____, _____, _____

22, _____, _____, _____, _____, _____, _____, _____, _____

35, _____, _____, _____, _____, _____, _____, _____, _____

52, _____, _____, _____, _____, _____, _____, _____, _____

1.NBT.1

2) Write the numbers to show the **hours** on the clock face.

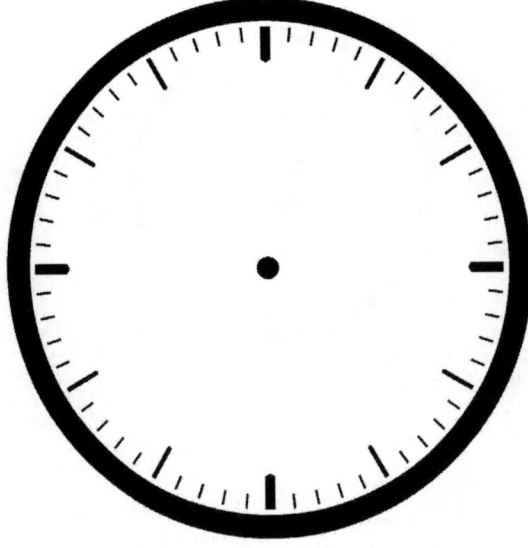

1.MD.3

Do you understand? ? ✓

Affirmation: I always try, even when I am scared.

Name: _____ Date: _____

Unit 7 Practice 16: I can draw hands on a clock face.

3) Draw the minute-hand to show the time.

11:30   4:00   6:00

1.MD.3

4) Draw the hands to show the time.

9:00   9:30

1.MD.3

Do you understand?   ?   ✓

Affirmation: Making mistakes is how I learn new things.

Name: _____ Date: _____

Unit 7 Practice 17: I can practice addition and subtraction. I can draw shapes.

1) Add or subtract to solve.

30 + 10 = ☐   60 - 10 = ☐

30 + 20 = ☐   60 - 30 = ☐

30 + 30 = ☐   45 - 15 = ☐

30 + 15 = ☐   30 - 15 = ☐

2.NBT.5

2) Draw one circle, one square, one triangle and one rectangle. **Label** each shape.

1.G.1

Do you understand?   ?   ✓

Affirmation: Making mistakes is how I learn new things.

Name: _____ Date: _____

Unit 7 Practice 17: I can practice addition and subtraction. I can draw shapes.

Questions A and B refer to the information below.

A) What time does the hour-hand-only clock show?

The clock shows [__ : __]

Explain your thinking.

B) Show **between**.
The hour hand on the clock is between the number _____ and the number _____.

Draw a circle between two rectangles.

1.MD.3

K.G.1

Do you understand?   ?   ✓

Affirmation: I get better the more I practice.

Name: _____ Date: _____

I can practice grade level fluencies.

Set 1: Find the sum.

1) 3 + 17 = _____

2) 12 + 7 = _____

3) 15 + 5 = _____

4) 8 + 11 = _____

5) 9 + 9 = _____

6) 9 + 8 = _____

7) 12 + 5 = _____

1.OA.6

Set 2: Find the difference.

1) 20 - 4 = _____

2) 17 - 7 = _____

3) 17 - 9 = _____

4) 13 - 3 = _____

5) 13 - 6 = _____

6) 16 - 5 = _____

7) 18 - 9 = _____

1.OA.6

Do you understand?   ?   ✓

Unit Reflection

Geometry and Time

Use this space to reflect on your understanding of the unit skills and concepts.

Skill/Concept	I can...	I need to work on...

Unit 8
Putting It All Together

In this unit we will practice adding and subtracting to become faster and more accurate. We will also solve different kinds of story problems and use place value to work with numbers up to 120.

Addition
23 + 12 = 34

Subtraction
34 - 12 = 23

Place Value

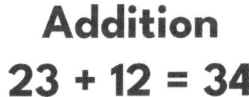

52 has 5 tens 2 ones

Compare Numbers

51 > 31
43 < 84
98 = 98

Story Problem:
Paul has 22 shapes.
Twelve of Paul's shapes are triangles.
The rest of Paul's shapes are circles.
How many shapes does Paul have?

12 + ☐ = 22

Paul has 10 circles.

Unit Vocabulary

Putting It All Together

Use this space to visualize the math vocabulary for this unit.

Word or Phrase	Example or Attributes	Visual Reminder

Unit Models & Strategies

Putting It All Together

Use this space to visualize the math models and strategies for this unit.

Model or Strategy	This is a...	It is used to...

Unit Models & Strategies

Affirmation: Making mistakes is how I learn new things.

Name: _____ Date: _____

Unit 8 Practice 1: I can find sums to ten.

1) Complete each number bond.

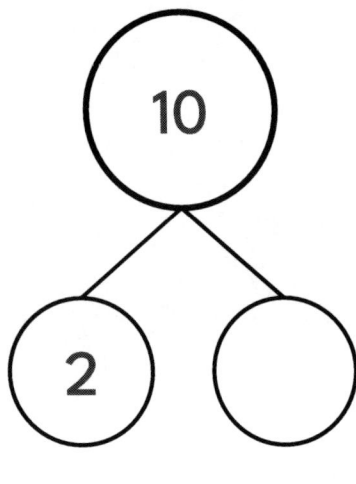 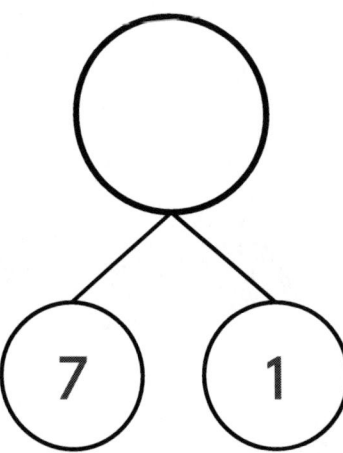 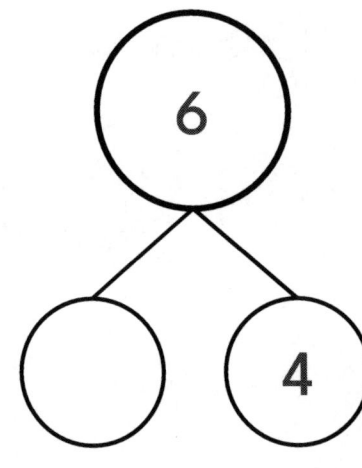

K.OA.3

2) Use the ten-frame to model 4 + 4.

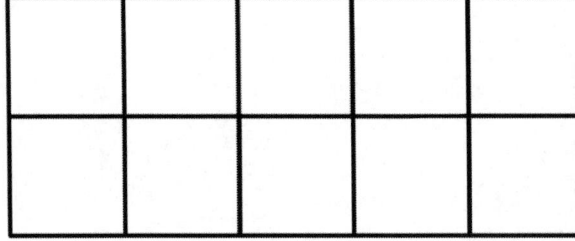

Explain your thinking.

Equation: _____

1.OA.6

Do you understand?   ?   ✓

Affirmation: Making mistakes is how I learn new things.

Name: _____ Date: _____

Unit 8 Practice 1: I can find sums to ten.

3) Counting on to solve.
   Draw a model to show 6 + 3.

   Explain your thinking.

   Equation: _____

   1.OA.5

4) Pam picked these cards.

   [10]   [1]

   What addition number puzzle could she make?

   Equation: _____

   1.OA.8

Do you understand?   ?   ✓

Affirmation: Making mistakes is how I learn new things.

Name: _____ Date: _____

Unit 8 Practice 2: I can relate addition and subtraction problems.

1) Solve each equation.
   Match the related addition and subtraction equations.

   $6 + 2 = \square$ $10 - \square = 1$

   $\square = 4 + 3$ $\square - 2 = 6$

   $1 + \square = 10$ $\square - 4 = 3$

   1.OA.8

2) Use the counters to model 9 - 4.

Explain your thinking.

Equation: _____

1.OA.6

Do you understand?   ?   ✓

Affirmation: Making mistakes is how I learn new things.

Name: _____ Date: _____

Unit 8 Practice 2: I can relate addition and subtraction problems.

3) Write an addition and a subtraction equation to match the counters in the ten-frame.

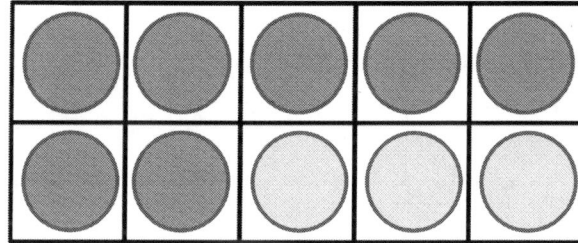

Addition equation: _____

Subtraction equation: _____

1.OA.4

4) Chloe says she can solve 9 - 2 using addition.
   Do you **agree** or **disagree** with Chloe?
   I _____ with Chloe.

   Show how you would solve 9 - 2.
   Explain your thinking.

1.OA.4

Do you understand?   ?   ✓

I Can Do Math Practice Problems, Grade 1

Affirmation: I am brilliant, bright, and getting better every day!

Name: _____ Date: _____

Unit 8 Practice 3: I can add and subtract to 20.

1) Subtract.
   Match the number that makes each equation true.

   18 - ☐ = 14   ☐ 11

   ☐ - 5 = 6   ☐ 17

   24 - 5 = ☐   ☐ 4

   ☐ = 20 - 3   ☐ 9

   1.OA.8

2) Use the double ten-frame to solve 20 - 9.

   Explain your thinking.

   Equation: _____

   1.OA.6

Do you understand? ? ✓

Affirmation: I am brilliant, bright, and getting better every day!

Name: _____   Date: _____

Unit 8 Practice 3: I can add and subtract to 20.

3) Show how you can make a ten to solve.

$$6 + 5 + 4$$

Equation: _____

1.OA.6

4) Use cube towers and cubes to solve.

$$3 + 6 + 7$$

Equation: _____

1.OA.6

Do you understand?   ?   ✓

Affirmation: I am brilliant, bright, and getting better every day!

Name: _____ Date: _____

Unit 8 Practice 4: I can solve addition and subtraction story problems.

1) Write the number that makes each equation true.

   10 - ☐ = 15   ☐ = 7 + 8

   ☐ - 8 = 3   13 - 2 = ☐

   9 + 5 = ☐   11 + ☐ = 17

   1.OA.8

2) There were 15 pigs splashing in the mud puddle.
   Some more pigs jumped into the mud puddle.
   Then there were 19 pigs playing in the mud puddle.
   How many pigs jumped into the mud puddle?

   Equation: _____

   1.OA.1

Do you understand?   ?   ✓

Affirmation: I am brilliant, bright, and getting better every day!

Name: _____   Date: _____

Unit 8 Practice 4: I can solve addition and subtraction story problems.

3) There were 20 cows in a field.
   Some cows walked back to the barn.
   Now there are 12 cows in the field.
   How many cows walked back to the barn?

   Equation: _____

   1.OA.1

4) Farmer Jones counted 13 animals in the barn.
   There were 4 horses and the rest were sheep.
   How many sheep were in the barn?

   Equation: _____

   1.OA.1

Do you understand?   ?   ✓

I Can Do Math Practice Problems, Grade 1

Affirmation: I am brilliant, bright, and getting better every day!

Name: _____ Date: _____

Unit 8 Practice 5: I can solve change unknown story problems.

1) Write the number that makes each equation true.

6 + ☐ = 10   12 + ☐ = 20

5 + ☐ = 10   4 + ☐ = 20

2 + ☐ = 10   10 + ☐ = 20

3 + ☐ = 10   7 + ☐ = 20

1.OA.4

2) Mr. Potter has a pet store.
There are 17 fish in one tank.
Eight are goldfish and the rest are clownfish.
How many clownfish are in the tank?

Equation: _____

1.OA.1

Do you understand?   ?   ✓

Affirmation: I am brilliant, bright, and getting better every day!

Name: _____ Date: _____

Unit 8 Practice 5: I can solve change unknown story problems.

3) Mr. Potter has 11 pets for adoption in his pet store.
   Six are cats and the rest are dogs.
   How many dogs does Mr. Potter have for adoption in his pet store?

   Equation: _____

   1.OA.1

4) Mr. Potter has turtles for sale.
   Four turtles are on a rock and some are in the water.
   There are 12 turtles for sale.
   How many turtles are in the water?

   Equation: _____

   1.OA.1

Do you understand?   ?   ✓

I Can Do Math Practice Problems, Grade 1

Name: _____ Date: _____

Unit 8 Practice 6: I can use two equations to solve compare story problems.

1) Circle true 👍 or false 👎 .

5 + 4 = 5 + 2 + 2
👍 👎

6 + 9 = 8 + 7
👍 👎

20 = 18 - 2
👍 👎

12 - 4 = 8 + 0
👍 👎

4 + 9 = 15 - 2
👍 👎

4 + 3 + 7 = 10 + 4
👍 👎

1.OA.7

2) Ellie saw 12 monkeys and 8 parrots at the zoo. How many more monkeys did she see than parrots?

Write two different equations to solve.

Equation: _____

Equation: _____

1.OA.4

Do you understand?   ?   ✓

Affirmation: I am brilliant, bright, and getting better every day!

Name: _____ Date: _____

Unit 8 Practice 6: I can use two equations to solve compare story problems.

3) Wyatt counted 6 giraffes during his visit to the zoo.
   Bruce counted 10 zebras during his visit to the zoo.
   How many fewer giraffes than zebras were counted during the visit?

   Write two different equations to solve.

   Equation: _____

   Equation: _____

   1.OA.4

4) Nora saw lions and tigers at the zoo.
   She drew this chart to compare the number of lions and tigers she saw.

   Lions [ ][ ][ ][ ][ ][ ][ ][ ][ ][ ]

   Tigers [ ][ ][ ][ ][ ][ ]

   How many more lions than tigers did Nora see?

   Equation: _____

   1.OA.4

Do you understand?   ?   ✓

I Can Do Math Practice Problems, Grade 1

Affirmation: I am brilliant, bright, and getting better every day!

Name: _____ Date: _____

Unit 8 Practice 7: I can model and count collections to 120.

1) Count on by ones to 120.

   97, _____, _____, _____, _____, _____, _____, _____, _____

   101, _____, _____, _____, _____, _____, _____, _____, _____

   111, _____, _____, _____, _____, _____, _____, _____, _____

   1.NBT.1

2) Hannah wants to draw 100 using base-ten blocks.
   She has drawn one ten.
   Finish Hannah's drawing.

   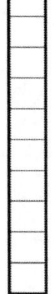

   1.NBT.1

Do you understand?   ?   ✓

Affirmation: I am brilliant, bright, and getting better every day!

Name: _____ Date: _____

Unit 8 Practice 7: I can model and count collections to 120.

3) Show 109 with base-ten blocks **or** with cube trains and cubes.

1.NBT.1

4) How many in this collection?

The number is _____.

1.NBT.1

Do you understand? ? ✓

Affirmation: I love thinking! I am a thinker.

Name: _____ Date: _____

Unit 8 Practice 8: I can count and draw models for two-digit numbers.

1) Eliana wants to model the number 31 with base-ten blocks.
   She has one ten and a bagful of ones.
   How can Eliana use one ten and a lot of ones to model 31?

   1.NBT.2

2) Model 27 in two different ways.

   1.NBT.2

Do you understand?   ?   ✓

Affirmation: I love thinking! I am a thinker.

Name: _____ Date: _____

Unit 8 Practice 8: I can count and draw models for two-digit numbers.

Questions A and B refer to the information below.

[base-ten blocks: 4 tens rods and 24 ones units]

A) Write the 2-digit number for the collection above.

The number is _____.

1.NBT.2

B) Draw base-ten blocks to show the collection in a different way.

1.NBT.2

Do you understand? ? ✓

I Can Do Math Practice Problems, Grade 1

Affirmation: I love thinking! I am a thinker.

Name: _____ Date: _____

Unit 8 Practice 9: I can solve number riddles.

1) Match the number that will make each true.

43 > ☐   27

24 < ☐   10

28 + 3 = 30 + ☐   39

47 − 7 = 30 + ☐   1

1.NBT.3

2) Tony is thinking of a number.
   The number is greater than 20.
   The number is less than 30.
   The number has 6 ones.
   What is Tony's number?

Tony's number is _____.

1.NBT.3

Do you understand?   ?   ✓

Affirmation: I love thinking! I am a thinker.

Name: _____ Date: _____

Unit 8 Practice 9: I can solve number riddles.

3) Choose a two-digit number that is greater than 34 and less than 43.
   Write your number.
   Draw a model of your number.

   My number is _____.

   1.NBT.3

4) Victoria is playing a cover-up game.
   She has a collection of base-ten blocks that equal 51.
   She has 24 base-ten blocks on the table and has the rest inside the box.
   How many base-ten blocks are inside the box?

   There are _____ base-ten blocks inside the box.

   1.NBT.4

Do you understand? ? ✓

Affirmation: I love thinking! I am a thinker.

Name: _____ Date: _____

Unit 8 Practice 10: I can use models to write number riddles.

1) Find the sum.
   Model 24 + 32 using base-ten blocks.

   Equation: _____

   1.NBT.4

2) Write <, >, or = to make each comparison true.

   10 + 3 ◯ 30 + 3   60 + 3 ◯ 70 - 7

   14 + 7 ◯ 20 + 0   12 + 26 ◯ 36

   4 + 90 ◯ 90 + 4   20 + 5 ◯ 15 + 10

   103   ◯ 50 + 53   8 + 10 ◯ 80 + 10

   1.NBT.3

Do you understand? ? ✓

Affirmation: I love thinking! I am a thinker.

Name: _____   Date: _____

Unit 8 Practice 10: I can use models to write number riddles.

3) Write the number.
   Match the collection of base-ten blocks to the true statements.

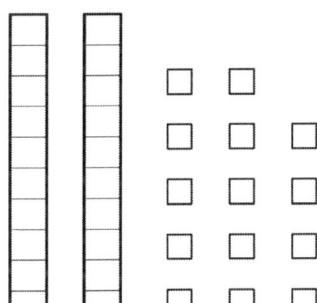

| 3 tens 4 ones |

| _____ > 40 |

| _____ < 40 |

| 2 tens 14 ones |

The number is _____.

1.NBT.2

4) Complete each clue to make it true.

67

The number is greater than _____.

The number is less than _____.

The number has _____ tens.

The number has _____ ones.

1.NBT.2

Do you understand?   ?   ✓

I Can Do Math Practice Problems, Grade 1

Affirmation: I get better the more I practice.

Name: _____ Date: _____

I can practice grade level fluencies.

Set 1: Add multiples of ten.

1) 53 + 40 = _____

2) 40 + 29 = _____

3) 12 + 70 = _____

4) 50 + 16 = _____

5) 38 + 30 = _____

6) 60 + 16 = _____

7) 10 + 72 = _____

1.NBT.4

Set 2: Subtract multiples of ten.

1) 45 - 20 = _____

2) 62 - 30 = _____

3) 19 - 10 = _____

4) 39 - 20 = _____

5) 100 - 80 = _____

6) 102 - 20 = _____

7) 100 - 60 = _____

1.NBT.6

Do you understand?   ?   ✓

Unit Reflection

Putting It All Together

Use this space to reflect on your understanding of the unit skills and concepts.

Skill/Concept	I can...	I need to work on...

Made in the USA
Middletown, DE
31 January 2026

27880711R00084